dna

A Bridge Between Biochemistry & Biotechnology

DNA
A Bridge Between Biochemistry & Biotechnology

Authors :
J.J. DHRUVE, M.Sc., Ph.D., (*Agril. Officer*)
U.K. KANDOLIYA, M.Sc., (*Agril. Officer*)
S.V. PATEL, M.Sc., Ph.D., (*Assoc. Res. Scientist*)
B.A. GOLAKIYA, M.Sc., Ph.D., (*Professor and Head*)

Department of Biochemistry, College of Agriculture
Junagadh Agricultural University, Junagadh
Gujarat, India

New India Publishing Agency
Pitam Pura, New Delhi- 110 088

Published by
Sumit Pal Jain *for*
New India Publishing Agency
101, Vikas Surya Plaza, CU Block, L.S.C. Mkt.,
Pitam Pura, New Delhi- 110 088, (India)
Phone: 011-27341717, Fax: 011-27341616
E-mail: newindiapublishingagency@gmail.com
Web: www.bookfactoryindia.com

ISBN 978-81-89422-24-0

Typeset at: Laxmi Art Creations, Delhi
Printed at: Jai Bharat Printing Press, Delhi

Dr. B. K. Kikani
Vice-Chancellor

Junagadh Agricultural University
Junagadh - 362 001

FOREWORD

In a way Bioscience is busy in understanding the cause and origin of life, the organization and the continuation of life and ultimately the development and sustenance of healthy and joyful life. Our understanding of primordial life penetrated up to DNA, its composition, transformation and transduction.

The secret of life is encoded in DNA. Our knowledge about this life string will make us wiser about the pleasure and pain of life. It will tell us so many things about the on going life on organism and a makeup of a living being yet to be borne through the existing organism.

It has taken several centuries of scientific efforts to arrive up to DNA. The history of DNA is noless absorbing than a class detective novel. Beside the reorus efforts of the scientist, it seems that nature was also vulnerable to disclose the secret of life and therefore it was possible to chart out the conceptional and actual frame and functionalities of an intricate mechanism operating in and through the DNA. The scientific information so generated about the DNA is voluminous. We wanted to narrate it in a simple and a lucid form. It will be palatable to kids and also to those who are least familiar to biosciences. Our purpose of so doing is to attract the people of diverse faculties towards this wonderful object.

Befriending the DNA will ventilate many unknown dimensions of life. Note only that, this wonder polymer has many applications and uses in the disciplines other than life science.

When our knowledge about any natural entities becomes holistic we can playfully implicate, imitate and manipulate its nature, properties and functions for the better life for all the living being around us.

As such this book has no any discipline biase. It will attract the larger section of people interested in the masterpiece controlling the life system on this planate. This collection entitled ***"DNA: A Bridge Between Biochemistry and Biotechnology"*** is an humbel effort of our team at the department of biochemistry towards popularizing bioscience to widen the horizons of knowledge.

Date : 25-04-2008
Place: Junagadh

(B.K. Kikani)

Dr. D.B. Kuchhadiya
Director of Research and
Dean P.G. Studies

Junagadh Agricultural University
Junagadh - 362 001

MESSAGE

Nature packs-up the secretes of life in microspace. The DNA is a micro-cosmos. Facination about the genetic map attracts even a common man. We wants to popularize the science of DNA up to kids level, because its concerns to everybody. This document *"DNA : A Bridge Between Biochemistry and Biotechnology"* is a versatile collection covering 12 chapters exclusively related to DNA. It starts with cell, cell division, chromosomes, taking deep dive in to structural and functional aspects of DNA and gene and gene expression. It also narrates working method with DNA, DNA-cloning and DNA libraries; enzyme dealing with DNA. The last chapter consists of the detail physio-chemical properties of the chemicals required in the working with DNA. The DNA glossary will be useful to all trying to understand the scientific aspects of DNA.

The getup of the text material is so attractive, adoned with graphs, figures and flow charts that any reader will enjoy the trip of this book. It has always been a tough task to make lucide and beautiful scientific document. Our team in the department of biochemistry and biotechnology is trying to do this hard task. Their sole purpose is to popularize the bioscience from kids to common man. I hope this document will serve that purpose.

(D.B. Kuchhadiya)

Date 25-04-2008
Place: Junagadh

PREFACE

The biological science are at the cross road. Evolution, propagation and expression of the traits of the organism confine to DNA - the wonder molecule embodied in the cell. The secret of this protein polymer has been revealed in the early 20th century. Now a days maximum time, energy and fund is being invested in understanding the structure and function of DNA. Possibly the answers to most of our questions related to life on this earth lies in the DNA. A lot of many methods to study the DNA from different facets have been invented. Our knowledge about this miraculous molecule is increasing every day. All engaged in the life science be the academicians or professionals wants to know the basics and advances taking place in understanding DNA.

The DNA is of concern for the pharma, food, fibre, plastics, energy, industry and even up to schooling guys. We decided to cater the need of the all of them in a single document. Thus, this collection has been composed to explain the history, chemistry, biochemistry and even research methods involved in understanding the DNA. A few protocols have also been described for the beginners working with DNA. The properties of the reagents used in working with DNA are described in the last chapter. It will be helpful in understanding the cause effect relations operating in the standard procedures involved in studying DNA. It will also help resolve the methodological artifacts encountered in the laboratory work related to DNA. The glossary of terms related to DNA is the last portion of this text. It will be useful in understanding any further reading related to DNA . We hope that this little book will help in promoting the science of DNA.

Authors

INTRODUCTION

DNA is a digitalized system of a nature, which is working on the base of more advance and complex 4 bites system *i.e.* "A", "T", "G" and "C" a four nitrogenous bases other than two bites "0" and "1" in man made machine. In all the living kinds 1 bites known as "Codon" composed of three bites from various combination of "A", "T", "G" and "C" is a basic unit of information or programming known as a "Genes" in a Nature. A British biologist, Richard Dawkins in his book "The selfish Gene" (1989) has said about the gene, "they are in you and me, they created us, body and minds and their presentation is the ultimate rationale for our existence..... they go by the name of genes and we as their survival machines".

Although, being a same genetic makeup of every cell in single organisms, functions are different according to tissues. All this things are because of complex nature of "DNA programming" and its implementation, expression in a genes of different cell or tissues. So, understanding a genomic DNA and its nature play an important role to working with biotechnological aspects. Recombinant DNA technology is sophisticated and expensive. Genes must be isolated, vectors must be identified, and gene control must be maintained. Stability of the vector within a host cell is important, and the scientist must be certain that nonpathogenic bacteria are used. Cells from mammals can be used to synthesize proteins, but cultivating these cells is difficult. In addition, the proper gene signals must be identified, RNA molecules must be bound to ribosomes, and the presence of introns must be considered. Collecting the gene product and exporting it from the cell are other considerations. So, this is an honest attempt to make a bridge/path between biochemistry and biotechnology via DNA.

What is Biochemistry?

- It is the "Chemistry of the living cell"
- Uses basic laws of chemistry, biology and physics to explain processes of living cells

Why Study Biochemistry?

- Lead us to fundamental understanding of life
- Understand important issues in medicine, health, and nutrition
- Has led to greater molecular understanding of diseases such as diabetes, sickle cell anemia, and cystic fibrosis.
- Next frontier: AIDS, cancer, Alzheimer's Disease

Advance Biotechnology Industries

- Biotechnology is the application of biological cells, cell components, and biological properties to technically and industrially useful operations.

Tools to Study Biochemistry

- Know chemical structures and reactivities of molecules that participate in cellular reactions
- Know biological function of cellular molecules
- Know how all of the pieces and different pathways fit together
- Use knowledge from general chemistry, organic chemistry, and biology and apply it to biological systems. Concepts and mechanisms are the same.

What is Biotechnology?

Biotechnology means any technological application that uses biological systems, living organisms, or derivatives thereof, to make or modify products or processes for specific use.

Why Study Biotechnology?

Biotechnology is an important tool that can improve our quality of life in many ways, both now and in the future – from life-saving medicines to more nutritious foods.

Tools to Study Biotechnology

The basic process of recombinant DNA technology revolves around DNA activity in the synthesis of protein. During this synthesis, DNA provides the genetic code for the placement of amino acids in proteins. By intervening in this process, scientists can change the nature of the DNA, thereby changing the nature of the protein expressed by that DNA. By inserting genes into the genome of an organism, the scientist can induce the organism to produce a protein it does not normally produce. The common tools used in biotechnology are as under.

- Antisense Technology
- Binding of DNA, Antigens, and Antibodies
- Electrophoresis, Blotting Techniques, and PCR
- DNA Microarrays
- Selectable Marker Genes
- Bioinformatics
- Genomics

CONTENTS

1

The Cell

All living organisms on earth are divided in pieces called cells. There are larger pieces called tissues and systems, while smaller pieces to cells that include organelles as well as proteins. Cells are small compartments that hold all of the biological equipment necessary to keep an organism alive and successful on the Earth. A main purpose of a cell is to organize. Cells hold a variety of pieces and each cell has a different set of functions. It is easier for an organism to grow and survive when cells are present. If you were only made of one cell, you would only be able to grow to a certain size. You don't find single cells that are as large as a cow. Also, if you were only one cell you couldn't have a nervous system, no muscles for movement, and using the internet would be out of the question. The trillions of cells in your body make your life possible.

1. All living creatures are made from one or more cells.
2. All cells are produced from previously existing cells i.e.(no spontaneous generation).
3. All cells appear to be descended from the first cell which existed about 4 billion years ago.
4. For a species to exist its reproductive cells must be potentially immortal.
5. Our bodies start from a single cell and contain about 100,000,000,000,000 (10^{14}) cells at maturity.

Living cells are divided into two types - prokaryotic and eukaryotic. This division is based on internal complexity. The following pages can provide graphic roadmaps to the organization of both of these cell types.

1. Prokaryotic cells are more primitive, small and without organelles: bacteria, cyanobacteria (blue-green algae).
2. Eukaryotic cells are more advanced, larger, contain organelles: all higher species (animals, plants, fungi, protozoa).

1.1 Prokaryotic

These cells are simple in structure, with no recognizable organelles. They have an outer cell wall that gives them shape. Just under the rigid cell wall is the more fluid cell membrane. The cytoplasm enclosed within the cell membrane does not exhibit much structure when viewed by electron microscopy.

In other words, small prokaryotic cells are simple but Fast

- Small cells such as bacteria divide fast (~ 20 min)- due to the high surface/volume ratio)
- Size: mycoplasmas : 0.1-1 micron dia.; other bacteria 1-10 micron dia.
- Have no nucleus: DNA less protected, mutates faster
- One compartment- like a chem lab with a single beaker
- Most have a rigid cell wall

May have a capsule outside the cell wall, pili (for attachment), flagella (for motion)

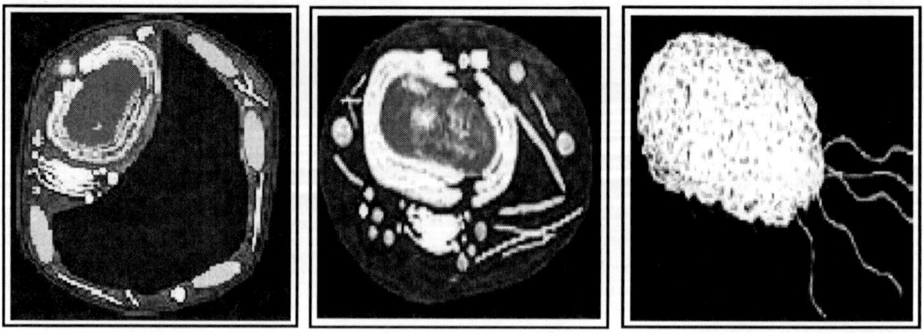

Plant Cell Animal Cell Bacterial Cell

Fig. 1.1. Plant, Animal and Bacteria Cell Models

1.2 Eukaryotic

The cells of protozoa, higher plants and animals are highly structured. These cells tend to be larger than the cells of bacteria, and have developed specialized packaging and transport mechanisms that may be necessary to support their larger size. Here are representative drawings of plant and animal cells showing their respective organelles.

Thus large eukaryotic cells are slow but versatile

- Cell division very slow (~ 20 hours)- due to the low surface/ volume ratio)
- Size: typically 10-100 micron dia.; volumes typically 1000 to 1 million times larger than prokaryotes
- Have nucleus: DNA better protected, slow mutation rate
- Have many organelles, most surrounded by bilayer lipid membranes
- Organelles allow many activities to take place within the same cell- like a chem lab with many beakers; other reactions take place on membrane surfaces and eukaryotic cells have much more internal membrane surface that prokaryotic cells
- Relative volumes of compartments in a liver cell:

1.3 Cell Orgenellie

The cell organelles are found within the cytosol

Cytosol is the liquid matrix of the cell- a watery solution containing a lot of protein, Contains salts, dissolved molecules, enzymes, etc. Glycolysis takes place in cytosol and gives 2 ATPs per glucose (does not require O_2), Cytoplasm = cytosol + organelles - nucleus.

Table 1.1 Cell Orgenellie

	The nucleus is the most obvious organelle in any eukaryotic cell. It is a membrane-bound organelle and is surrounded by a double membrane. It communicates with the surrounding cytosol via numerous nuclear pores. Within the nucleus is the DNA responsible for providing the cell with its unique characteristics. The DNA is similar in every cell of the body, but depending on the specific cell type, some genes may be turned on or off - that's why a liver cell is

Contd...

Table 1.1. Contd...

Nucleus and Nucleolus 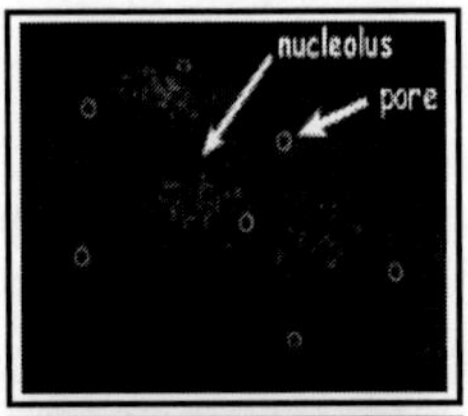 	different from a muscle cell, and a muscle cell is different from a fat cell. When a cell is dividing, the DNA and surrounding protein condense into chromosomes that are visible by microscopy. The prominent structure in the nucleus is the nucleolus. The nucleolus produces ribosomes, which move out of the nucleus to positions on the rough endoplasmic reticulum where they are critical in protein synthesis.
Endoplasmic Reticulum 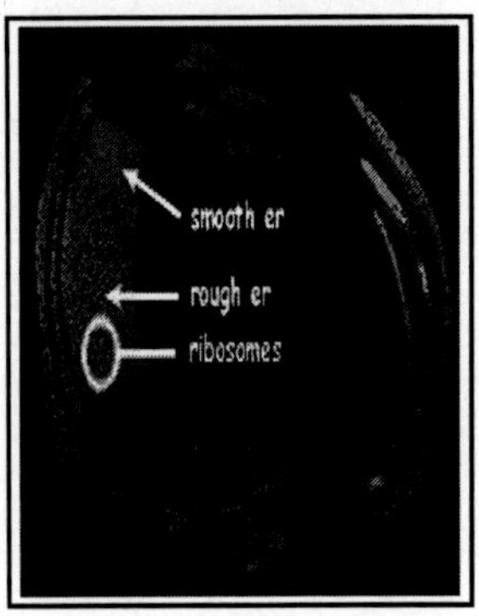	Throughout the eukaryotic cell, especially those responsible for the production of hormones and other secretory products, is a vast amount of membrane called the endoplasmic reticulum (ER). The ER membrane is a continuation of the outer nuclear membrane and its function suggests just how complex and organized the eukaryotic cell really is. When viewed by electron microscopy, the appearance of ER divide in two parts i.e. some areas look "smooth" (smooth ER) and some appear "rough" (rough ER). The rough ER appears rough due to the presence of ribosomes on the membrane surface. Smooth and Rough ER also have different functions. Smooth ER is important in the synthesis of lipids and membrane proteins. Rough ER is important in the synthesis of other proteins. Information coded in DNA sequences in the nucleus is transcribed as messenger RNA. Messenger RNA exits the nucleus through small pores to enter the cytoplasm. At the ribosomes on the rough ER, the messenger RNA is translated into proteins. These proteins are then transferred to the Golgi in "transport vesicles" where they are further processed and packaged into lysosomes, peroxisomes, or secretory vesicles.
	The centrosome, also called the "microtubule organizing center", is an area in the cell where

Contd...

Table 1.1. Contd...

The Centrosome and the Centrioles 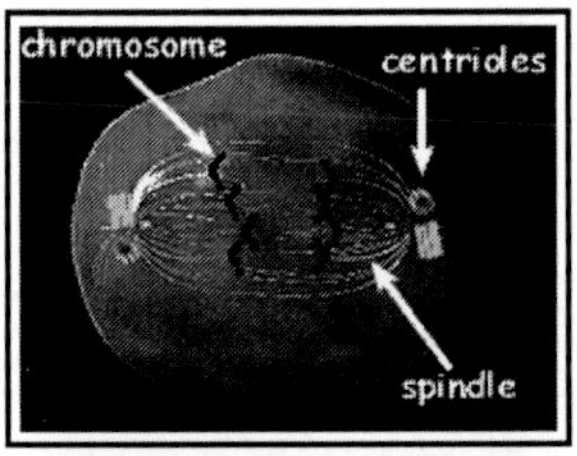 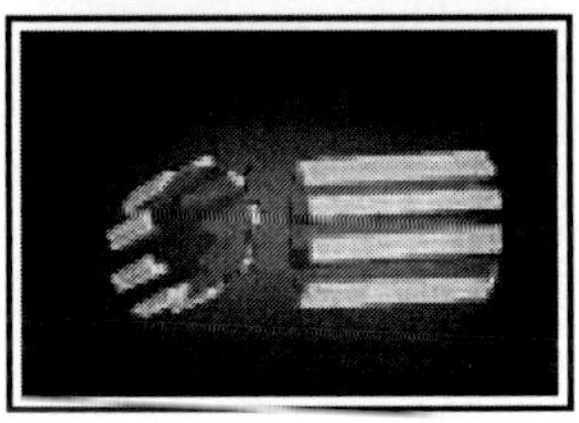	microtubles are produced. Within an animal cell centrosome there is a pair of small organelles, the centrioles, each made up of a ring of nine groups of microtubules. There are three fused microtubules in each group. The two centrioles are arranged such that one is perpendicular to the other. During animal cell division, the centrosome divides and the centrioles replicate. The result is two centrosomes, each with its own pair of centrioles. The two centrosomes move to opposite ends of the nucleus, and from each centrosome, microtubules grow into a "spindle" which is responsible for separating replicated chromosomes into the two daughter cells. Plant cell chromosome: Plant cells have centrosomes that function much like animal cell centrosomes. However, unlike centrosomes in animal cells, they do not have centrioles.
Golgi Apparatus 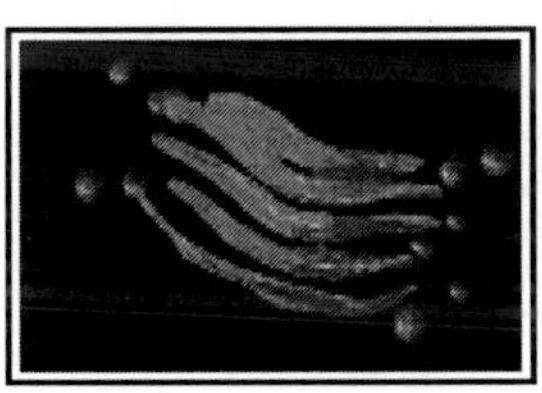	The Golgi apparatus is a membrane-bound structure with a single membrane. It is actually a stack of membrane-bound vesicles that are important in packaging macromolecules for transport elsewhere in the cell. The stack of larger vesicles is surrounded by numerous smaller vesicles containing those packaged macromolecules. The enzymatic or hormonal contents of lysosomes, peroxisomes and secretory vesicles are packaged in membrane-bound vesicles at the periphery of the Golgi apparatus.
Cytoskeleton 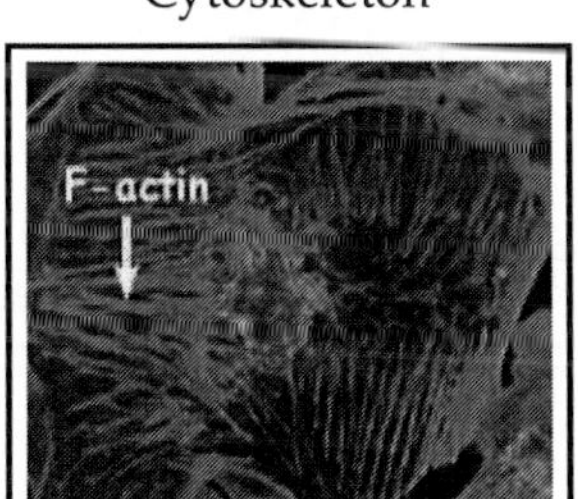	As its name implies, the cytoskeleton helps to maintain cell shape. But the primary importance of the cytoskeleton is in cell motility. The internal movement of cell organelles, as well as cell locomotion and muscle fiber contraction could not take place without the cytoskeleton. The cytoskeleton is an organized network of three primary protein filaments: microtubules, actin

Contd...

Table 1.1. Contd...

	filaments, and intermediate fibers. The complexity of the cytoskeleton can be seen in the abundant F-actin stress fibers (green) in the endothelial cell shown above.
Cell Membrane 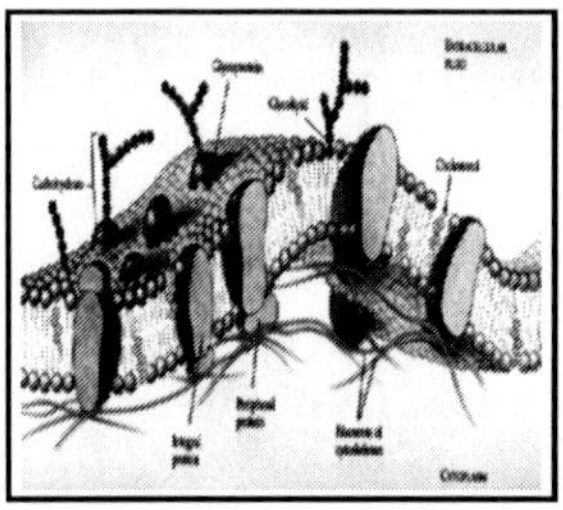	Every cell is enclosed in a membrane. The membrane is a double layer of lipids (lipid bilayer) but is made quite complex by the presence of numerous proteins that are important to cell activity. These proteins include receptors, pores, and enzymes. The membrane is responsible for the controlled entry and exit of ions like sodium (Na) potassium (K), calcium (Ca++).
Cytosol 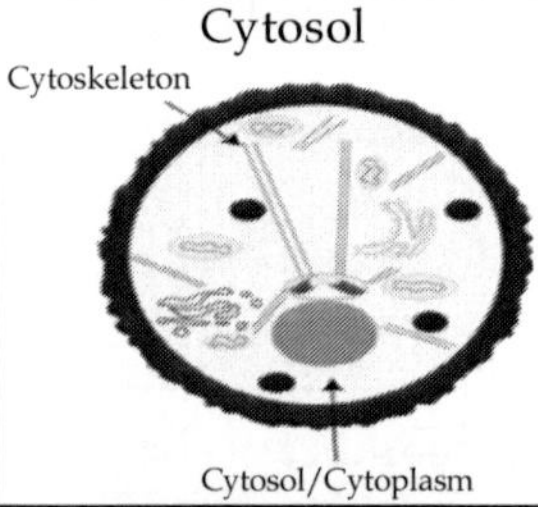	The cytosol (cytoplasm) is the "soup" within which all the other cell organelles reside and where most of the cellular metabolism occurs. Though mostly water, the cytosol is full of proteins that control cell metabolism including signal transduction pathways, glycolysis, intracellular receptors, and transcription factors
Mitochondria 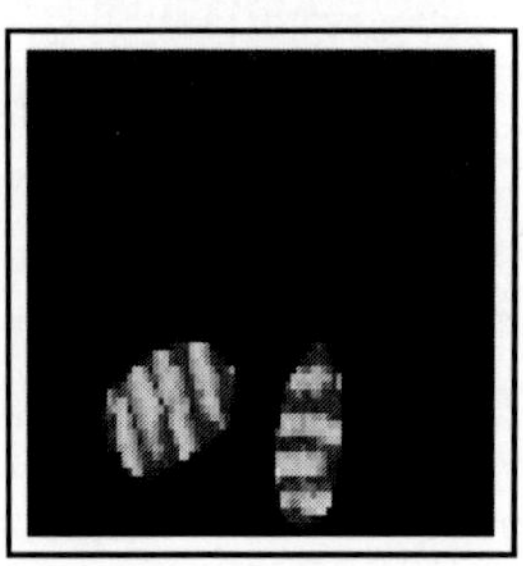	Mitochondria provide the energy a cell needs to move, divide, produce secretory products, contract - in short, they are the power centers of the cell. They are about the size of bacteria but may have different shapes depending on the cell type. Mitochondria are membrane-bound organelles, and like the nucleus have a double membrane. The outer membrane is fairly smooth. But the inner membrane is highly convoluted, forming folds called cristae. The cristae greatly increase the inner membrane's surface area. It is on these cristae that food (sugar) is combined with oxygen to produce ATP - the primary energy source for the cell.
	Chloroplasts are specialized organelles found in all higher plant cells. These organelles contain

Contd...

Table 1.1. Contd...

Chloroplast 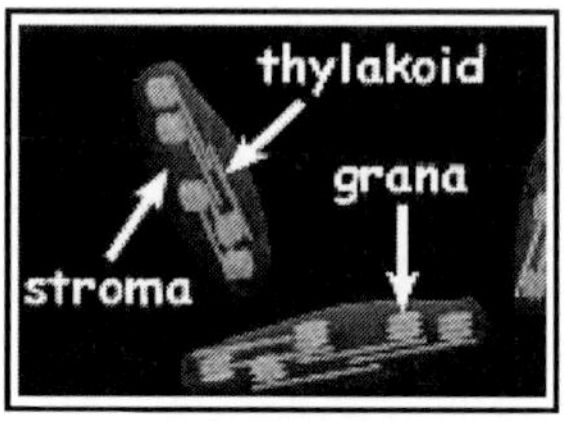	the plant cell's chlorophyll, hence provide the green color. They have a double outer membrane. Within the stroma are other membrane structures - the thylakoids and grana (singular = granum) where photosynthesis takes place.
Lysosomes 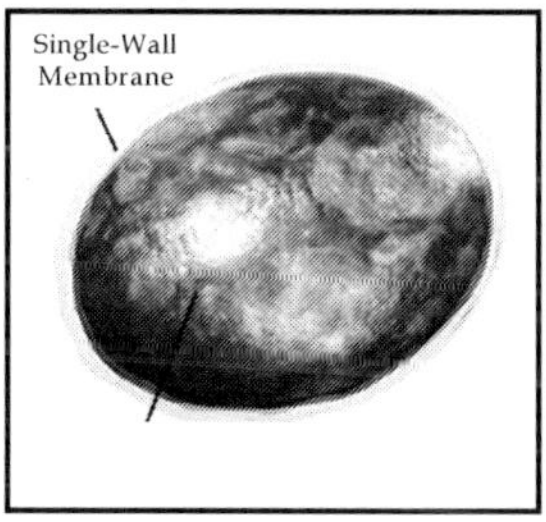	Lysosomes (common in animal cells but rare in plant cells) contain hydrolytic enzymes necessary for intracellular digestion. Lysosomes contain digestive enzymes that break down proteins, lipids, etc. It can also break down defective cell parts so they can be recycled. In white blood cells that eat bacteria, lysosome contents are carefully released into the vacuole around the bacteria and serve to kill and digest those bacteria. Uncontrolled release of lysosome contents into the cytoplasm can also cause cell death (necrosis)
Peroxisomes 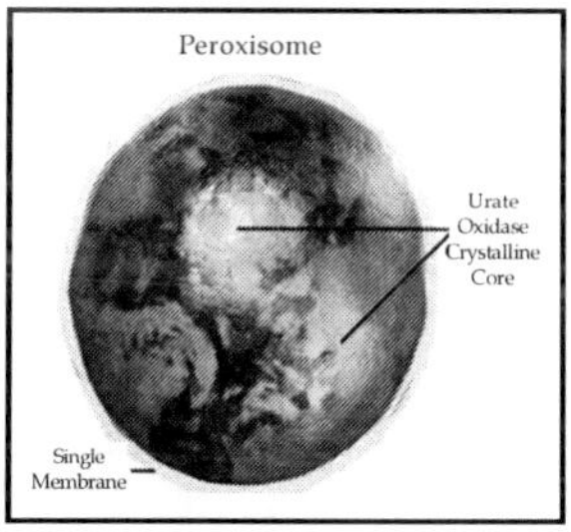	This organelle is responsible for protecting the cell from its own production of toxic hydrogen peroxide. It contain the enzyme, catalase, which converts hydrogen peroxide to O_2 and water. As an example, white blood cells produce hydrogen peroxide to kill bacteria. The oxidative enzymes in peroxisomes break down the hydrogen peroxide into water and oxygen. It can most Important in fat metabolism.
Secretory Vesicles 	Cell secretions - e.g. hormones, neurotransmitters - are packaged in secretory vesicles at the Golgi apparatus. The secretory vesicles are then transported to the cell surface for release
	A vacuole is a membrane-bound sac that plays roles in intracellular digestion and the release of

Contd...

Table 1.1. Contd...

Vacuole 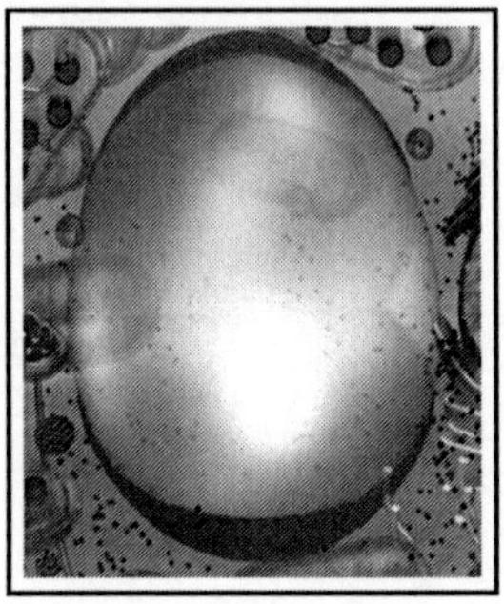	cellular waste products. Cells contain small vacuoles ("containers"), used for storing material such as glycogen and lipids. Some protozoa (i.e., ameoba) have a contractile vacuole. When the cell takes in too much water (osmosis) the vacuole lining contracts, squirting out excess water. In animal cells, vacuoles are generally small. Vacuoles tend to be large in plant cells and play a role in turgor pressure. When a plant is well-watered, water collects in cell vacuoles producing rigidity in the plant. Without sufficient water, pressure in the vacuole is reduced and the plant wilts.
Cell Wall 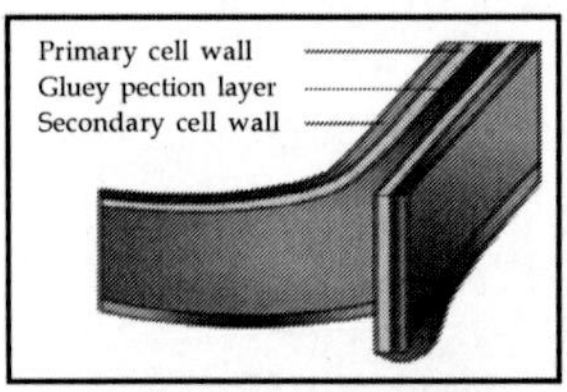	Prokaryotic cells and plant cells both have a rigid cell wall made up of polysaccharides. The cell wall provides and maintains the shape of these cells and serves as a protective barrier. Has little effect on cell permeability but gives rigidity to cells. Like an external skeleton for plant cells. Gives protection against osmotic lysis.

1.2 Differences between prokaryotes and Eukaryotes

	Prokaryotes	**Eukaryotes**
Organism type	Bacteria, blue green algae	Animals, Plants, fungi
Form of organisms	Mostly unicellular, a few are linear filaments of unicells	Some are unicellular but most are multi cellular with a complex division of labour between different types of cells
Cell Size	About 1 –10 m	About 10 – 100 m
Location of genetic material	DNA occurs as a single circular chromosome which is not membrane bound	Chromosomes are located inside an envelope formed from two membranes i.e. nucleus. There are pores in the envelope
Other structures in cytoplasm	Some membranes may exist but never forming an envelope	All contain mitochondria, others depending on species, contain chloroplasts, liposome's, golgibody, endoplasmic reticulm
Ribosomes	These small particles are involved in protein synthesis. In prokaryotes ribosomes are relatively small.	Ribosomes are relatively large
Metabolism	Some are autotrophos and some are heterotrophos. Of the heterotrophs, some are aerobic and some are anaerobic.	All plants are autotrophos. Of the heterotrophs most are aerobic. Some can be anaerobic.

Some differences between Plant cells and Animal cells are as follow:

- Plant cells have cell walls as their outermost layer.
- Plant cells have chloroplasts that contain chlorophyll for pigmentation
- Plant cells have larger vacuoles (part of the endomembrane system used for storage) than Animal cells
- Animal cells contain centrioles that play a role in mitosis
- Animal cells have flagellum connected to the cell membrane which aids in movement of the cell.

❑❑❑

2

THE CELL CYCLE

The sequence of activities exhibited by cells is called the cell cycle. This sequence is appeared during development from stem to fully differentiate, cells in the body alternately divide (mitosis) and "appear" to be resting (interphase). The cell cycle is an ordered set of events, culminating in cell growth and division into two daughter cells. Non-dividing cells not considered being in the cell cycle. The stages (Fig.2.1), are G1-S-G2-M. The G1 stage stands for "GAP 1". The S stage stands for "Synthesis". This is the stage when DNA replication occurs. The G2 stage stands for "GAP 2". The M stage stands for "mitosis", and is when nuclear (chromosomes separate) and cytoplasmic (cytokinesis) division occur. Mitosis is further divided into 4 phases *i.e.* prophase, metaphase, anaphase and telophase.

2.1 Regulation of the cell cycle

Cell division (and thus tissue growth) is controlled is very complex. The following terms are some of the features that are important in regulation, and places where errors can lead to cancer. Cancer is a disease where regulation of the cell cycle goes awry and normal cell growth and behavior is lost.

Cdk (cyclin dependent kinase, adds phosphate to a protein), along with cyclins, are major control switches for the cell cycle, causing the cell to move from G1 to S or G2 to M.

MPF (Maturation Promoting Factor) includes the CdK and cyclins that triggers progression through the cell cycle.

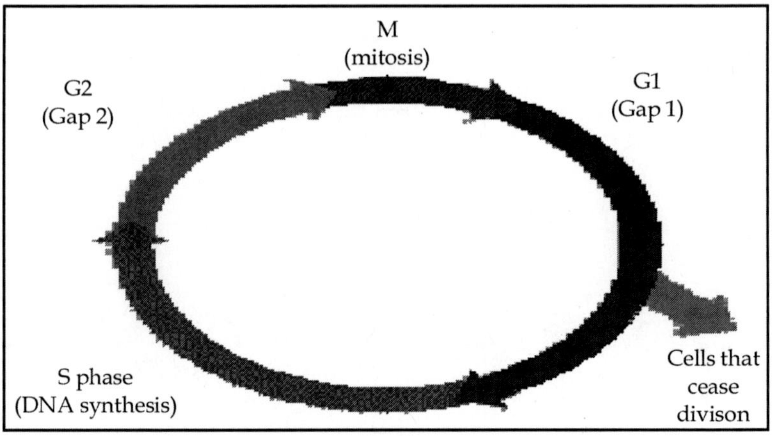

Fig. 2.1. The cell cycle

p53 is a protein that functions to block the cell cycle if the DNA is damaged. If the damage is severe this protein can cause apoptosis (cell death).

1. p53 levels are increased in damaged cells. This allows time to repair DNA by blocking the cell cycle.
2. A p53 mutation is the most frequent mutation leading to cancer. An extreme case of this is Li Fraumeni syndrome, where a genetic a defect in p53 leads to a high frequency of cancer in affected individuals.

2.2 Mitosis

Mitosis is nuclear division plus cytokinesis, and produces two identical daughter cells during prophase, prometaphase, metaphase, anaphase, and telophase. Interphase is often included in discussions of mitosis, but interphase is technically not part of mitosis, but rather encompasses stages G1, S, and G2 of the cell cycle.

Interphase

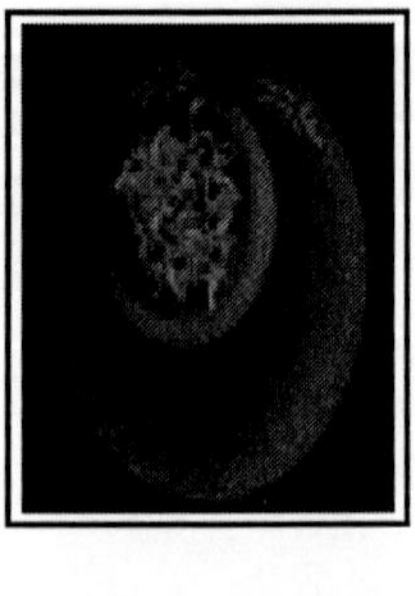

Cells may appear inactive during this stage, but they are quite the opposite The cell is engaged in metabolic activity and performing its prepare for mitosis (the next four phases that lead up to and include nuclear division). Chromosomes are not clearly discerned in the nucleus, although a dark spot called the nucleolus may be visible. The cell may contain a pair of centrioles (or microtubule organizing centers in plants) both of which are organizational sites for microtubules. This is the longest period of the complete cell cycle during which DNA replicates, the centrioles divide, and proteins are actively produced. For a complete description of the events during Interphase.

Prophase

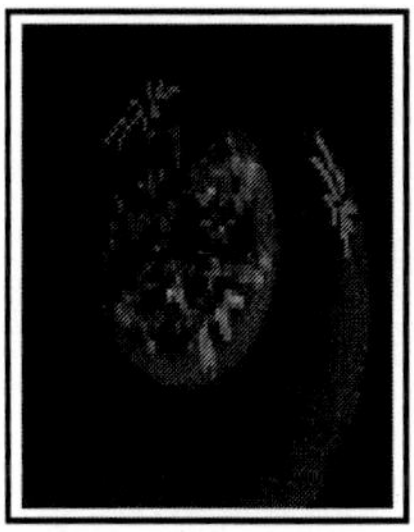

During this first mitotic stage, the nucleolus fades and chromatin (replicated DNA and associated proteins) Chromatin in the nucleus begins to condense and becomes visible in the light microscope as chromosomes. The nucleolus disappears. Centrioles begin moving to opposite ends of the cell and fibers extend from the centromeres. Some fibers cross the cell to form the mitotic spindle Each replicated chromosome comprises two chromatids, both with the same genetic information. Microtubules of the cytoskeleton, responsible for cell shape, motility and attachment to other cells during interphase, disassemble. And the building blocks of these microtubules are used to grow the mitotic spindle from the region of the centrioles.

Prometaphase

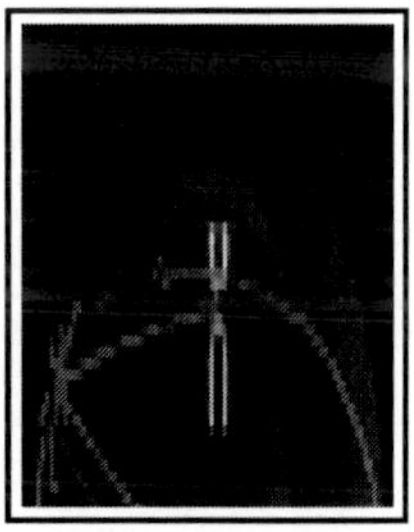

In this stage the nuclear envelope breaks down so there is no longer a recognizable nucleus The nuclear membrane dissolves, marking the beginning of prometaphase. Some mitotic spindle fibers elongate from the centrioles and attach to kinetichores, protein bundles located on the chromosomes. Proteins attach to the centromeres creating the kinetochores Microtubules attach at the kinetochores and the chromosomes begin moving Other spindle fibers elongate but instead of attaching to chromosomes, overlap each other at the cell center.

Metaphase

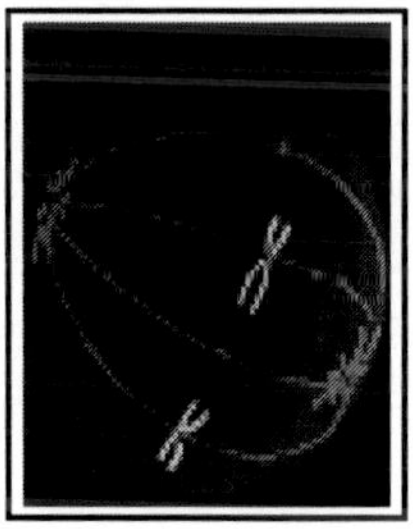

Spindle fibers align the chromosomes along the middle of the cell nucleus. This line is referred to as the metaphase plate. Tension applied by the spindle fibers aligns all chromosomes in one plane at the center of the cell. This organization helps to ensure that in the next phase, when the chromosomes are separated, each new nucleus will receive one copy of each chromosome.

Anaphase

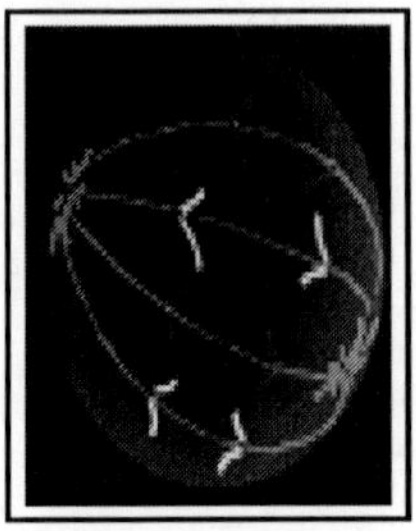

The paired chromosomes separate at the kinetochores and move to opposite sides of the cell. Spindle fibers shorten, the kinetichores separate, and the chromatids are pulled apart and begin moving to the cell poles. Motion results from a combination of kinetochore movement along the spindle microtubules and through the physical interaction of polar microtubules.

Telophase

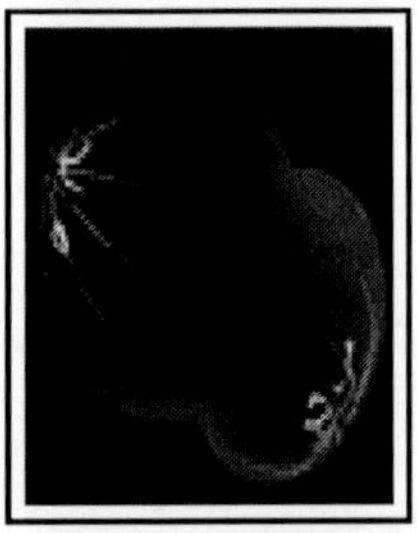

Chromatids arrive at opposite poles of cell, and new membranes form around the daughter nuclei. The chromosomes disperse and are no longer visible under the light microscope. The spindle fibers disperse, and cytokinesis or the partitioning of the cell may also begin during this stage.

Cytokinesis

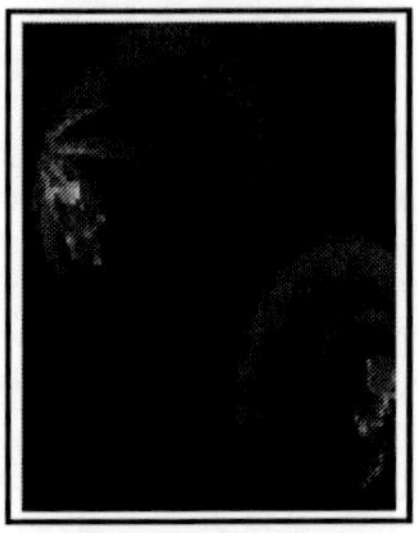

The spindle fibers not attached to chromosomes begin breaking down until only that portion of overlap is left. It is in this region that a contractile ring cleaves the cell into two daughter cells. Microtubules then reorganize into a new cytoskeleton for the return to interphase. In animal cells, cytokinesis results when a fiber ring composed of a protein called actin around the center of the cell contracts pinching the cell into two daughter cells, each with one nucleus. In plant cells, the rigid wall requires that a cell plate be synthesized between the two daughter cells.

2.3 Meiosis

There are two cell divisions. Mitosis has one division and meiosis has two divisions. Now here we still have to remember PMATI, but now we do it twice. We also need to remember that four cells are created where there was originally one. That's four cells with half of the amount of DNA needed by a cell. When a cell goes through meiosis, it's not concerned about creating another working cell.

Meiosis happens when it's time to reproduce an organism. The steps of meiosis are very simple. When we break it down it's just two PMATI's in a row. Scientists say Meiosis I and Meiosis II, but it's just two PMATIs. The interphase that happens between the two processes is very short and the DNA is not duplicated.

As we said, meiosis happens when it's time to reproduce. Meiosis is the great process that shuffles the cell's genes around. Plants do it, animals do it, and even sometimes fungi also do it. Instead of creating two new cells with equal numbers of chromosomes (like mitosis), the cell does a second division soon after the first.

That second division divides the number of chromosomes in half. When you have half the number of chromosomes, you are called a haploid cell. Haploid means half the regular number. Diploid is the opposite (two strands). Normal cells are considered to be diploid cells.

Meiosis I

This is basically like the PMATI of a regular mitosis. Pairs of chromosomes are lined up at the center of the cell and then pulled to each side. Meiosis is a bit different because there something called crossing-over happens with the DNA.

The crossing over is an exchange of genes. These genes are mixed up, not resulting in a perfect duplicate like mitosis. Now the cell divides, and make leaving two new cells with a pair of chromosomes each. Normally the cell would begin to go about its business of living and slowly duplicate the chromosomes for another mitotic division. Since this is meiosis, there is a very short interphase and division begins again.

Meiosis II

In Prophase II the DNA that remains in the cell begins to condense and form short chromosomes. Each chromosome pair has a centromere. The centrioles also begin their journey to opposite sides of the cell. In Metaphase II all of the chromosomes line up along the center of the cell and the centrioles are in position for the duplication. Anaphase II shows the chromosomes split and move to opposite sides of the cell. Each one splits into two pieces. They don't divide up the DNA between the new cells; they split the DNA that exists. Each daughter cell will get one-half of the DNA needed to make a functioning cell.

Telophase II shows the DNA completely pulled to the sides and the cell membrane begins to pinch. When it's all over, you are left with four haploid cells that are called gametes. The eventual purpose of the gametes will be to find other gametes with which they can combine. When they do, they will form a new organism.

Significance of Meiosis

Meiosis is a mechanism in which the traits of both the parents are combined. The crossingover event before gamete formation allows an exchange of chromosomal genes between maternal and paternal chromosomes, to provide genetic variation population. While, the necessity of mitosis in sexually reproducing species ensures restoration of diploid (2n) number of chromosomes in an individual.

□□□

3

DNA: As Genetic Material and It's History

DNA (deoxyribonucleic acid), the giant molecule that carries genetic information in living things, is made up of just a few chemical building blocks that bond together in very particular ways. A typical molecule of DNA consists of two strands that are linked together. A segment can be visualized as a ladder-like structure:

The DNA as made of two strands, a left and a right, which are linked in the middle. In reality, since the ladder is twisted, there really is no "right" or "left," but we use that terminology here since it applies to the model graphs. The square shapes and diamonds you see in the middles of the "rungs" are the links, which are meant to represent the different chemical bonds between the two strands of DNA.

Each of the two strands of DNA consists of a sequence of nitrogenous bases, and any one of four such bases can appear on a strand. These molecules, our genes, are long polymers of deoxyribonucleic acid, or DNA. Just four chemical building blocks—guanine (G), adenine (A), thymine (T), and cytosine (C)—are placed in a unique order to code for all of the genes in all living organisms. The research about the DNA is represented as below:

Year	Scientist(s)/ Scientist groups	Events discover
1858	Charles Darwin Alfred and Russel Wallace	Joint announcement of the theory of natural selection-that members of a population who are better adapted to the environment survive and pass on their traits.
1859	Charles Darwin	"Species change through natural selection"
1865	Gregor Mendal	Heredity is transmitted indiscrete units.
1869	Friedrich Miescher	First DNA isolation, He has isolated a material rich in phosphorus from the cells and called it nuclein.
1879	Walter Flemming	He described chromosome behaviour during animal cell division. He has the given steps such as prophase, metaphase and anaphase.
1900	Hugo de Vries, Carl Correns and Erich von Tschermak	They expand awareness of the mendelian of inheritance in the world.
1902	Archibald Garrod	He pointed out the disease alkaptonuria is inherited according to Mendelian rules.
	Walter Sutton	The segregation pattern of chromosomes during meosis matched the segregation pattern
1905	Nettie Stevens and Edmund Wilson	Independently described the behavior of sex chromosomes-XX determines female; XY determines male.
1908	Archibald Garrod	Proposed that some human diseases are due to "inborn errors of metabolism" that result from the lack of a specific enzyme.
1909	Wilhelm Johannsen	Coined the word "Gene". He made the distinction between the phenotype and genotype.
1911	Thomas Hunt Morgan	Genes strung on chromosomes are the units of heredity.

Contd....

1927	Hermann J. Muller	Used x-rays to cause artificial gene mutations in Drosophila
1928	Fred Griffith	Proposed that some unknown "principle" had transformed the harmless R strain of Diplococcus to the virulent S strain
1931	Harriet B. Creighton and Barbara McClintock	Demonstrated the cytological proof for crossing-over in maize.
1941	George Beadle and Edward Tatum	They given the hypothesis "One gene One enzyme"
1944	Oswald Avery, Colin MacLeod and Maclyn McCarty	They showed that DNA can transform the properties of cells, clarifying the chemical nature of genes.
1944	Barbara Mc Clintock	Genes can jump around on chromosomes showing that the genome is more dynamic then previously thought.
1945	Max Delbruck	Organized a phage course at Cold Spring Harbor Laboratory which was taught for 26 consecutive years. This course was the training ground of the first two generations of molecular biologists
1950	Erwin Chargaff	Discovered a one-to-one ratio of adenine to thymine and guanine to cytosine in DNA samples from a variety of organisms.
1951	Rosalind Franklin	Obtained sharp X-ray diffraction photographs of DNA.
1952	Alfred Hershey and Martha Chase	Only the DNA of virus needs to enter a bacterium to infect it.
1953	Francis Crick and James Watson	The double helical structure of DNA
1955	Joe Hin Tjio	46 number of chromosome
	Arther Karnberg and his colleagues	Isolated DNA polymerase an enzyme later used for all kinds of recombinant DNA technology and for sequencing.

Contd....

1956	Vernon Ingram	Specific chemical alteration in a heamoglobin protein- the substitution of valine for glutamic acid in the sixth amino acid in beta globin is the root of the disease.
1958	Matthew Meselson and Frank Stahl	DNA replicates semiconservatively with each strand in a DNA molecule from the parent generation ending up paired with a new strand in the daughter generation. Each parent strand of DNA served as a templete for the synthesis of a new strand of DNA.
1958	Arthur Kornberg	Purified DNA polymerase I from E. coli, the first enzyme that made DNA in a test tube.
1959	Jerome Lejeune and his colleague	Chromosome abnormalities identified
1961	Sydney Brener, Francois Jacob and Mathew Meselson	MRNA is the molecule that takes information from DNA in the nucleus to the protein making machinery in the cytoplasm.
1961	Robert Guthrie	First screen for metabolic defect in new borns.
1966	Marshall Nirenberg, H. Gobind Khorana and Severo Ochoa	Elucidated the genetic code.
1968	M. Meselon's group, H.O. Smith, K.W. Wilcox and T.J. Velly	First restriction enzyme described
1970	Hamilton Smith and Kent Wilcox	Isolated the first restriction enzyme, HindII, that could cut DNA molecules within specific recognition sites.
1972	San Franscio and Stanford and other researchers	First recombinant DNA
1973	Stanford and USCI researchers	First amino acid gene cloned

Contd....

1973	Joseph Sambrook	Led the team at Cold Spring Harbor Laboratory that refined DNA electrophoresis by using agarose gel and staining with ethidium bromide.
1973	Annie Chang and Stanley Cohen	Showed that a recombinant DNA molecule can be maintained and replicated in E. coli.
1975-77	Sanger and his colleagues Maxam and Gillbert	Developed rapid DNA sequencing methods.
1976	Genetech	The first genetic engineering co.
1977	Richard Roberts and Phil Sharp	Discovered introns
1981-82		First transgenic mice and fruit flies
1982		GeneBank data base formed
1983		First disease gene mapped
1983	James Gusella	Used blood samples collected by Nancy Wexler and her co-workers to demonstrate that the Huntington's disease gene is on chromosome 4.
1983		PCR invented
1985	Kary B. Mullis	Published a paper describing the polymerase chain reaction (PCR), the most sensitive assay for DNA yet devised.
1986		First time disease gene positionally cloned.
1987		First human genetic map (Based on RFLPs)
1987		YACs (Yeast Artificial Chromosomes) developed
1988		The Human Genome Project began with the goal of determining the entire sequence of DNA composing human chromosomes

Contd....

1989		Microsatellites, a new genetic markers develop which is useful tool as a genetic land marks to distinguish people).
1989	Alec Jeffreys	Coined the term DNA fingerprinting and was the first to use DNA polymorphisms in paternity, immigration, and murder cases.
1989	Francis Collins and Lap-Chee Tsui	Identified the gene coding for the cystic fibrosis transmembrane conductance regulator protein (CFTR) on chromosome 7 that, when mutant, causes cystic fibrosis.
1989		Sequence- tagged sites (STS) another marker developed
1990		Launch the human genome project DOE Department of Energy; NIH – National Institute of Health
1990		ELSI (Ethical, Legal and Social Implications) found by DOE and NIH
1991		ESTs (An expressed sequence tag) fragments of genes
1992		Second generation map of human genome
1992		Data release guidelines established
1993		New Human Genome Project established for five years
1994		Detailed human genetic map
1995		Two microbial genome sequenced
1995		Physical map of human genome completed
1996		• First international strategy meeting on human genome sequencing • Mouce genetic map completed • Yeast genome sequenced • Archea genome sequenced

Contd....

		• Health insurance discrimination banned • 280000 expressed sequence tags • Human gene map created
1996		Human DNA sequencing begins (National Human Genome Research Institute)
1997		E. Coli genome sequenced Recommendations on genetic testing by NIH and DOE.
1998		• Private company Celera Genomic announces sequencing plan • M. Tuberculosis bacterium sequenced • SACGT (Secretary of Health and Human Services Advisory Committee on Genetic Testing) formed • HGP map includes 30,000 human gene • SNP (Single Nucleotide Polymorphism) initiative begins • Round worm C. elegans sequenced
1999		HGP Full scale sequencing begins Full length sequence of human chromosomes
2000		• Genetic discrimination in Federal workplace banned • Free access to genome information • Fruit fly genome sequenced • 21 human chromosomes appears to have less than 300 genes
2003		Pannel completion of human sequencing Mouse working draft
2004		Animal viruses found to encode mi RNAs
2004		Rat and Chicken Genomes Sequenced

Contd....

2004		FDA Approves First Microarray
2004		Refined Analysis of Complete Human Genome Sequence
2004		Surgeon General Stresses Importance of Family History
2005		miRNAs shown to act as oncogenes
2005		Chimpanzee Genomes Sequenced
2005		Hap Map Project Completed
2005		Trypanosomatid Genomes Sequenced
2005		Dog Genomes Sequenced
2006		Nat-siRNA discovered in plant
2006		The Cancer Genome Atlas (TCGA) Project Started
2006		Second Non-human Primate Genome is Sequenced
2006		Initiatives to Establish the Genetic and Environmental Causes of Common Diseases Launched
Future		The meeting food need of Asia's increasingly urbanized population requires increase in agricultural productivity and matching these increase to dietary changes and rising incomes. New understanding of plant as well as animal genes may offer ways to increase crop yields to the levels required to adequately and sustainably feed the growing population in Asia. As with biotechnology the challenge will be to maintain a healthy balance to avoid the potential changes arising from advances in genetics to put term instead to better use.

❏❏❏

4

DNA STRUCTURE

4.1 DNA: The word means

D	Deoxyribo	The pentose does not have any oxygen in position 2. Compare a deoxyribose with a ribose
N	Nucleic	These molecules were first found in the nucleus of the cell , before being found in mitochondria, chloroplasts (of plant cells), and in the cytoplasm of prokaryotes.
A	Acid	Only two of the three acid groups of the phosphoric acid are used to form the DNA chain. The third one gives the phosphoribo-backbone an acidic property.

Where it Founds: It is found in Nucleus of Eukaryotes and Nucleosome in Prokaryotes.

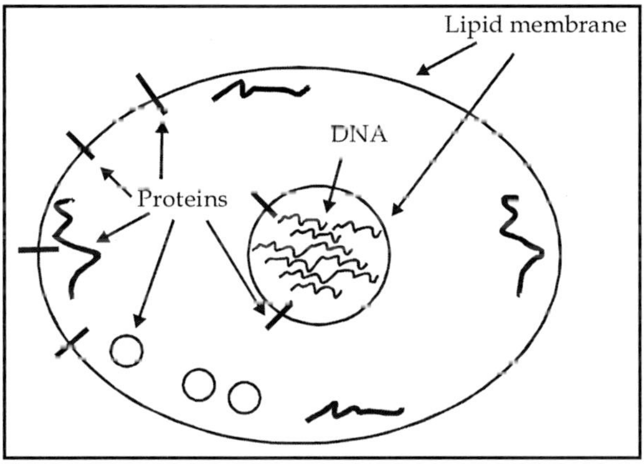

Fig: 4.1. Location of DNA in Cell

The deoxyribonucleic acid (DNA), is a long chain of nucleotides which consist of

1. Organic (nitrogenous) bases, (*Purines* - Adenine and Guanine *or* *Pyrimidines* -Cytosine and Thymine)
2. Deoxyribose (a pentose = sugar with 5 carbons)
3. Phosphoric Acid

Base **:** DNA (deoxyribonucleic acid) and RNA (ribonucleic acid) are composed of two different classes of nitrogen containing bases:

1. *Purines*: The most commonly occurring purines in DNA are adenine and guanine:
2. *Pyrimidins*: The most commonly occurring pyrimidines in DNA are cytosine and thymine:

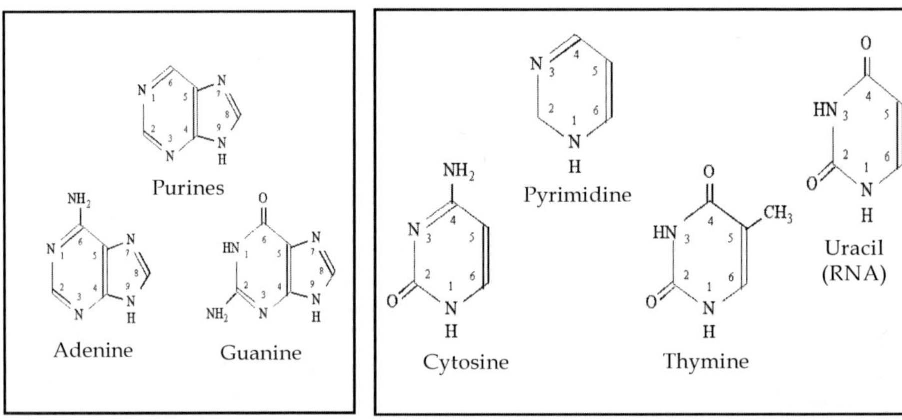

Fig. 4.2: Structure of Nitrogenus bases of Nucleotide Purine and Pyrimidine

RNA contains the same bases as DNA with the exception of thymine. Instead, RNA contains the pyrimidine uracil.

Sugar: In DNA and RNA the sugar found are pentose sugar, Ribose or Deoxyribose. Purines and pyrimidines bases can form chemical linkages with pentose (5-carbon) sugars. The carbon atoms on the sugars are designated 1', 2', 3', 4' and 5'. It is the 1' carbon of the sugar that becomes bonded to the nitrogen atom at position N1 of a pyrimidine or N9 of a purine. DNA precursors contain the pentose deoxyribose. RNA precursors contain the pentose ribose (which contains an additional OH group at the 2' position):

The resulting molecules are called nucleosides and can serve as elementary precursors for DNA and RNA synthesis, *in vivo*.

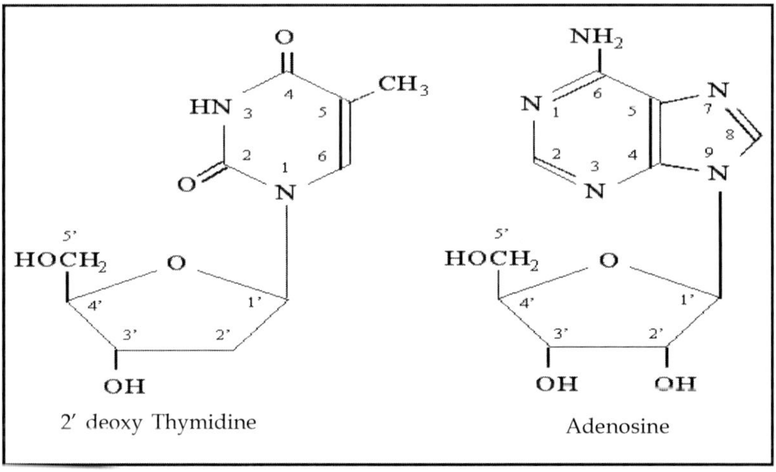

Fig. 4.3: Nucleoside structure

Phosphate: Before a nucleoside can become part of a DNA or RNA molecule it must become complexed with a phosphate group to form a nucleotide. Nucleotides can posess 1, 2 or 3 phosphate groups, e.g. the nucleotides adenosine monophosphate (AMP), adenosine diphosphate (ADP) and adenosine triphosphate (ATP). The phosphate groups are attached to the 5' carbon of the ribose sugar moiety. Beginning with the phosphate group attached to the 5' ribose carbon, they are labeled a, b and g phosphate. It is the tri-phosphate nucleotide which is incorporated into DNA or RNA.

Table 4.1: Base Components of DNA

Base	**Nucleoside**	**Nucleotide**	**RNA (monophosphate)**	**DNA (monophosphate)**	**Code**
Adenine	Adenosine	(Adenylic acid)	AMP	dAMP	A
Guanine	Guanosine	(Guanylic acid)	GMP	dGMP	G
Cytosine	Cytidine	(Cytidylic acid)	CMP	dCMP	C
Thymine	Thymidine	(Thymidylic acid)	TMP	dTMP	T
Uracil	Uridine	(Uridylic acid)	UMP	dUMP	U

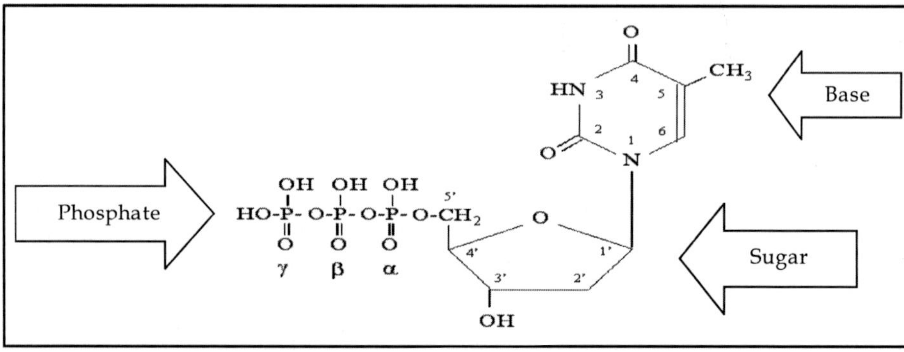

Fig: 4.4. A typical structure of Nucleotide

4.2 Double Helical Structure: The Watson and Crick model

DNA consists of two associated polynucleotide strands that wind together in a helical fashion. It is often described as a double helix. In a nucleotide, the atoms of the organic base are numbered 1, 2, ... and the atoms of the sugar, wether it is a deoxyribose like in DNA or a ribose like in RNA, are numbered 1', 2', 5'. Atoms in the sugar component of a nucleotide provide the link between the base and the phosphate group. The 1' carbon is attached to the 9 nitrogen of a purine, or the 1 nitrogen of a pyrimidine. The OH (hydroxyl) group on the 5' carbon is replaced by a bond to the phosphate group (ester bond). Each polynucleotide is a linear polymer in which the monomers (deoxynucleotides), are linked together by means of phosphodiester bridges, or bonds. These bonds link the 3' carbon in the ribose of one deoxynucleotide to the 5' carbon in the ribose of the adjacent deoxynucleotide.

The Watson and Crick model had the following dimensions

1. If G always paired with C, and T always paired with A, then either strand could be regenerated from the complementary information in the other strand.
2. The basis of the complementarity was hydrogen bonding, i.e. non-covalent interactions which could be easily broken and re-formed.
3. The information which DNA carried was within the unique base sequence of the DNA.
4. From the general interior location of the bases, it would appear that the double helix would have to dissociate in order to access the information.

5. The non-equitorial location of the sugar moieties (see above) suggested that the DNA helix would have a major groove and a minor groove.
6. The sugar / phosphate backbone is on the outside while the organic bases project into the inside of the double helix.
7. The organic bases stack on top of each other in parallel plans.
8. By convention a polynucleotide is read from the 5' end to the 3' end.
9. The orientations of the two strands are antiparallel : their 5' - 3' directions are opposite.
10. 34 Å per helical repeat
11. 10 base pairs per repeat (i.e. per turn of the helix)
12. 3.4 Å inter-base stacking distance
13. 20 Å diameter for the helical width
14. The two strands are held together by the energy of many hydrogen bonds and hydrophobic interaction. The base pairing is regular and precise.

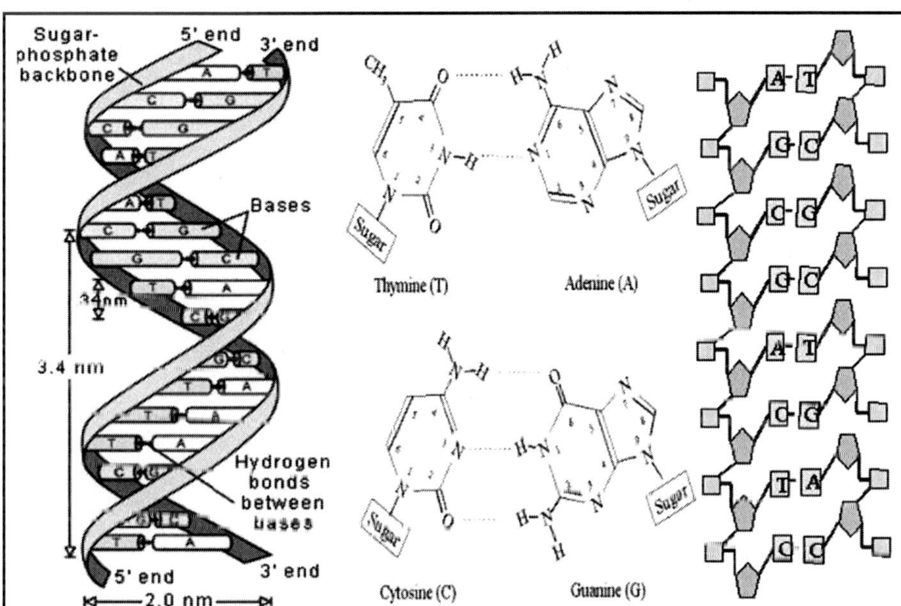

Fig. 4.5: Double helicle structure, hydrogen bonding and base pairing of DNA

The Hydrogen Bonding

In the "double helix" model of Watson and Crick the polynucleotide chains interact to form a double helix with the chains running in opposite directions. The bases are directed towards the center (and stack on top of one another) and the sugar backbones face the outside of the helix. The important property contributing to DNA helix stability is the stacking of the base-pairs on top of one another, due to hydrophobic forces. Free bases will stack on top of each other and form a helix in solution! This type of process is called "self-assembly", where you just throw something in solution, and it fits together on its own, with no extra work needed.

4.3 Family of DNA

A-DNA family : this is most common for double stranded RNA, RNA/DNA hybrids, as well as for certain DNA sequences, such as long stretches of purines.

B-DNA family : DNA exists in the "B-DNA form", most of the time inside the cells of living organisms. This is the classical "Watson-Crick" structure.

Z-DNA family : this is much more rare than the other two families, although certain sequences (such as runs of GC repeats (GCGCGC)) can form Z-DNA easily.

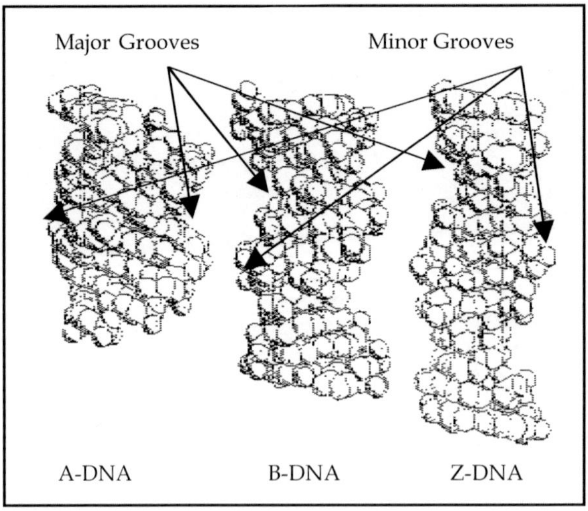

Fig 4.6: Type of DNA Family

The structure proposed by Watson and Crick is also called as B-DNA. Now many important deviations from the structure given by Watson and Crick have been found named A and Z-DNA. The A from is favored in a solution devoid of water. in his form, DNA is in the right handed double helix, but the rise per base pair is 0.23 nm compared to 0.34 nm in the B form and 11 nucleotides compared to 10 nucleotides per turn in the be from the z-form of the DNA is a left handed helical, having 12 nd: per turn of a helix and a rise of 0.38 nm per base pair. It looks likes a zig-zag structure.

4.4 The Chromosomes

In living cells, DNA molecules are always protected and packaged into complex, multidimensional structures that become "colored bodies" when stained with dyes and observed under a light microscope. From this property to take up dyes, we get the general term for these structures; chromosomes, a compound word derived from the words "chroma" meaning color, and "soma" meaning body.

The DNA molecule is twisted onto itself, and the supercoiled molecule is wrapped around proteins known as histone proten which maintain its shape. These proteins have a special name - histone proteins. The fibre has the structure of a spring (a solenoid, to be technical), as shown in the illustration on the right. The DNA is wrapped around the protein, and then, twisted together again in the shape of a spring. The complex DNA protein is called chromatin, while the structure formed by two turns of DNA around one histone is called a nucleosome. The chromosome fibre is coiled around a protein scaffold. The string of little packets is then wrapped around another protein structure to form a chromatid. A chromosome like this one is made of two identical molecules of DNA efficiently packed into chromatids. As can be seen in the illustration, a chromatid is made of a protein frame, and loops around it. The efficiency of this packing is pretty impressive! Human chromosome 1 is hardly 10 micrometers long. But it contains 7 centimetres of DNA.

A chromosome is the visible state of genetic material during a phase of the division of the cell (metaphase). Humans have 23 pairs of chromosomes, which makes the *diploid* number 46. The diploid number is the number of chromosomes of a normal cell. The *haploid* number is the number of chromosomes in a gamete. Usually, the diploid number is twice the haploid number. The number of chromosomes varies depending on the species. Just before mitosis, DNA, which contains the genetic information, is replicated.

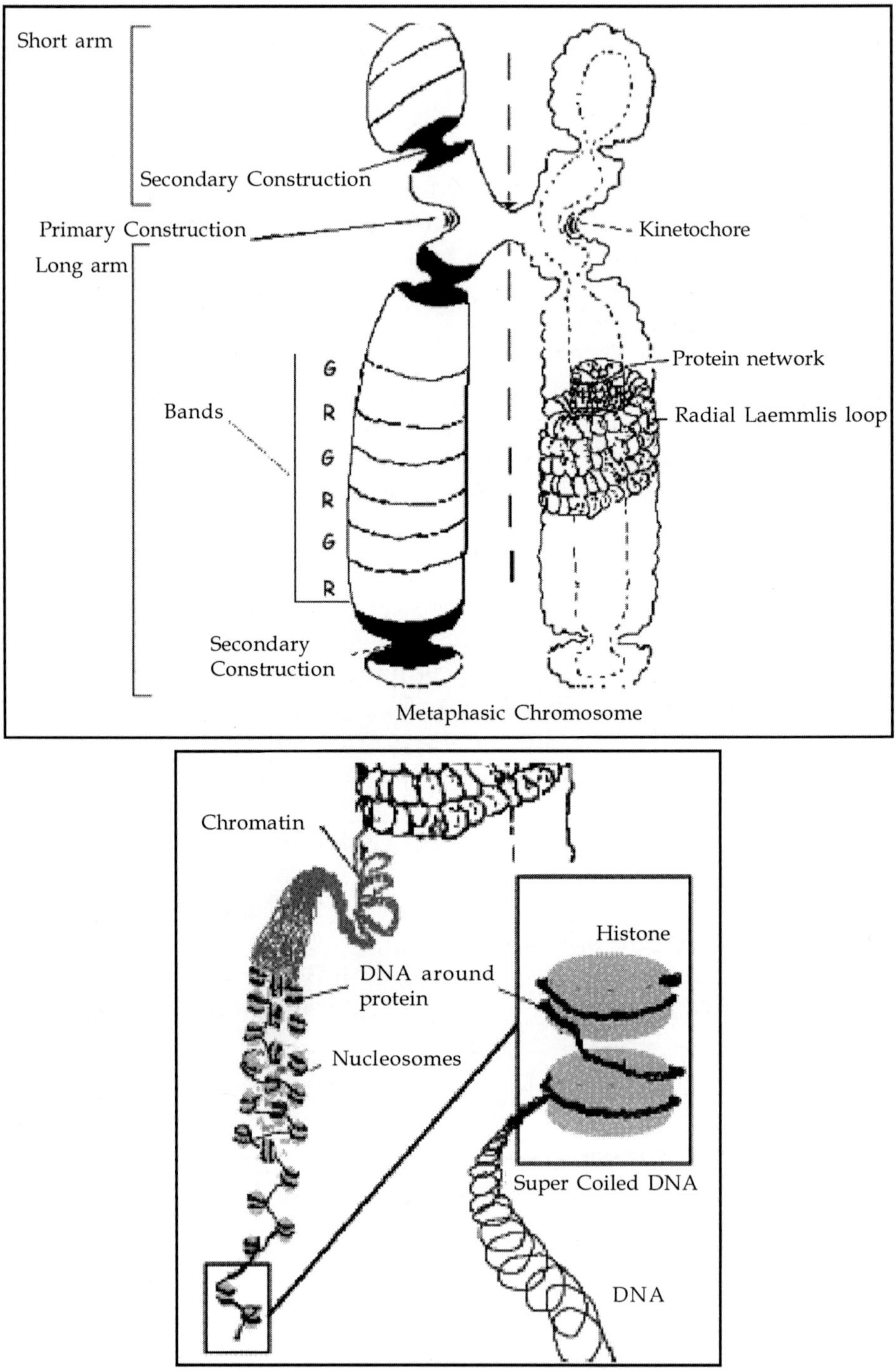

Fig. 4.7: A typical arrangement of DNA in Chromosome

DNA is a polynucleotide having deoxyribonucleotides as its constituents. In higher cells DNA is localized mainly in the nucleus as part of the chromosomes. A small amount of DNA presents in the cytoplasm and found within mitochondria and chloroplast. Now we see about the story of DNA i.e. when and how different scientists discovered this miracle of GOD.

The patterns of DNA by x-ray analysis its depending on the different conditions. Some important types of DNA are (1) Z- DNA (2) Noncoding DNA (3) Genomic DNA (4) Unique sequences DNA (5) Dispersed repetive DNA and (6) Satellite DNA.

(1) Z- DNA

- The helix is left handed
- Structure formed as Z shape
- The significance unknown

(2) Noncoding DNA

- In eukaryotic cells does not useful for code to make protein
- Genes have large intron sequences
- Genes are separated by stretches of sequence of unknown function

(3) Genomic DNA

- In bacteris the genome is a single or circular loop
- In eukaryotic cells it is the entire haploid nuclear complement of DNA
- In bacteria and the DNA of plasmids are Non genomic

(4) Uniquc sequences DNA

- The fraction of DNA is reassociates slowly is unique sequence
- E.Coli DNA is essentially all unique sequence

(5) Dispersed repetitive DNA

- It comprises a no. of DNA sequences of a few hundread base pairs
- In human, the ALU and the LI elements may replicating themselves by transposition

(6) Satellite DNA

- It occurs near the centromeres of chromosomes
- It involved in attachment of the mitotic spindle, consists of huge no. of tandem repeats of short sequences.

(7) A DNA

- It occurs in the low humidity atmosphere
- The helix is right handed
- The base pairs are titled and pulled away from the axis of the helix

(8) B DNA

- It occurs in the high humidity atmosphere
- The helix is right handed
- The helix axis passes through the base pairs
- The base pairs are oriented perpendicular to that axis

❑❑❑

5

DNA Replication Transcription and Translation

5.1 Replication

The cell in a plant synthesizes proteins and contains the appropriate DNA encoding for that information. All the cells are derived from a single cell in the very large stages of development, and then DNA has to be replicated many millions of times in order for all the cells to receive their complement.

DNA replication is the process in which a copy of DNA molecule is made and thus the genetic information it contains is duplicated. The parental double stranded DNA molecule is replicated semi conservatively i.e. each copy contains one of the original strands paired with a newly synthesized strand which is complementary in terms of AT and GC base pairing and of opposite chemical polarity. This process is very simple, however mechanically complex and involves multiple enzymatic activities.

The whole molecule may be replicated into a new molecule in which the old DNA remains the same and the new DNA is made of all new components, this process known as conservative replication. While, when the DNA broken up and new bits are added in a random pattern in which both strands of both resulting DNA molecules contain old and new components known as dispersive replication.

The mechanics of DNA replication was originally characterized in the bacterium, *E. coli* which contains 3 distinct enzymes capable of catalyzing the replication of DNA. These have been identified as DNA polymerase (pol) I, II, and III. Pol I is the most abundant replicating activity in *E. coli* but has as its primary role to ensure the fidelity of replication through the repair of damaged and mismatched DNA. Replication of the *E. coli* genome is the job of pol III. This enzyme is much less abundant than pol I, however, its activity is nearly 100 times that of pol I.

There have been 5 distinct eukaryotic DNA polymerases identified, a, b, g, d and e. The identity of the individual enzymes relates to its subcellular localization as well as its primary replicative activity. The polymerase of eukaryotic cells that is the equivalent of *E. coli* pol III is pol-a. The pol I equivalent in eukaryotes is pol-ß. Polymerase-g is responsible for replication of mitochondrial DNA.

The ability of DNA polymerases to replicate DNA requires a number of additional accessory proteins. The combination of polymerases with several of the accessory proteins yields an activity identified as DNA polymerase holoenzyme. These accessory proteins include followings:

- Primase
- Processivity accessory proteins
- Single strand binding proteins
- Helicase
- DNA ligase
- Topoisomerases
- Uracil-DNA N-glycosylase

The process of DNA replication begins at specific sites in the chromosomes termed origins of replication, requires a primer bearing a free 3'→OH, proceeds specifically in the 5'→3' direction on both strands of DNA concurrently and results in the copying of the template strands in a semiconservative manner. The semiconservative nature of DNA replication means that the newly synthesized daughter strands remain associated with their respective parental template strands.

The large size of eukaryotic chromosomes and the limits of nucleotide incorporation during DNA synthesis, make it necessary for multiple origins of replication to exist in order to complete replication in a reasonable period of time. A replication origin the strands of DNA must dissociate and unwind in order to allow access to DNA polymerase. Unwinding of the duplex at the origin as well as along the strands as the replication process proceeds is carried out by helicases. The binding of single-strand binding proteins stabilizes the resultant regions of single-stranded DNA. The stabilized single-stranded regions are then accessible to the enzymatic activities required for replication to proceed. The site of the unwound template strands is termed the replication fork.

In order for DNA polymerases to synthesize DNA they must encounter a free 3'-OH which is the substrate for attachment of the 5'-phosphate of the incoming nucleotide. During repair of damaged DNA the 3' OH can arise from the hydrolysis of the backbone of one of the two strands. During replication the 3'-OH is supplied through the use of an RNA primer, synthesized by the primase activity. The primase utilizes the DNA strands as templates and synthesizes a short stretch of RNA generating a primer for DNA polymerase.

Synthesis of DNA proceeds in the 5'→3′ direction through the attachment of the 5'-phosphate of an incoming dNTP to the existing 3'-OH in the elongating DNA strands with the concomitant release of pyrophosphate. Initiation of synthesis, at origins of replication, occurs simultaneously on both strands of DNA, synthesis then proceeds bidirectionally, with one strand in each direction being copied continuously and one strand in each direction being copied discontinuously. During the process of DNA polymerases incorporating dNTPs into DNA in the 5'→3' direction they are moving in the 3'→5' direction with respect to the template strand. In order for DNA synthesis to occur simultaneously on both template strands as well as bidirectionally one strand appears to be synthesized in the 3'→5' direction. Here one strand of newly synthesized DNA is produced discontinuously.

The strand of DNA synthesized continuously is termed the leading strand and the discontinuous strand is termed the lagging strand. The lagging strand of DNA is composed of short stretches of RNA primer plus newly synthesized DNA approximately 100-200 bases long. The lagging strands of DNA are also called Okazaki fragments. The concept of continuous strand synthesis is somewhat of a misnomer since DNA polymerases do not remain associated with a template strand indefinitely. The ability of a particular polymerase to remain associated with the template strand is termed its' processivity. The longer it associates the higher the processivity of the enzyme. DNA polymerase processivity is enhanced by additional protein activities of the replisome identified as processivity accessory proteins.

A model has been proposed where DNA polymerases exist as dimers associated with the other necessary proteins at the replication fork and identified as the replisome. The template for the lagging strand is temporarily looped through the replisome such that the DNA polymerases are moving along both strands in the $3' \rightarrow 5'$ direction simultaneously for short distances, the distance of an Okazaki fragment. As the replication forks progress along the template strands the newly synthesized daughter strands and parental template strands reform a DNA double helix. It means that only a small stretch of the template duplex is single-stranded at any given time.

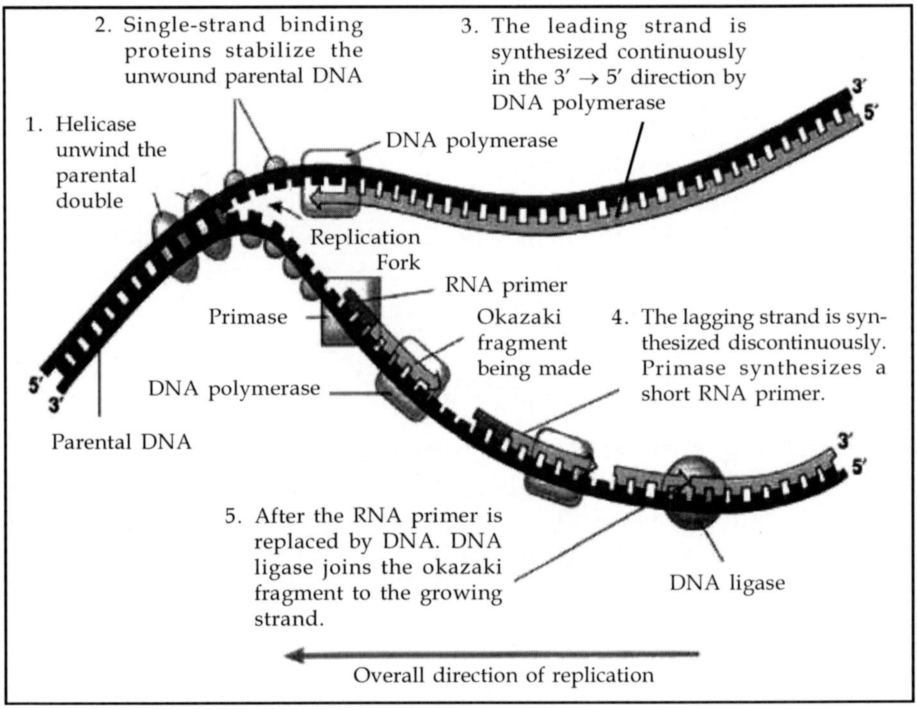

Fig. 5.1. Overview of Replication

The progression of the replication fork requires that the DNA ahead of the fork be continuously unwound. Due to the fact that eukaryotic chromosomal DNA is attached to a protein scaffold the progressive movement of the replication fork introduces severe torsional stress into the duplex ahead of the fork. This torsional stress is relieved by DNA topoisomerases. Topoisomerases relieve torsional stresses in duplexes of DNA by introducing either double- (topoisomerases II) or single-stranded (topoisomerases I) breaks into the backbone of the DNA. These breaks allow unwinding of the duplex and removal of the replication-induced torsional strain. The nicks are then resealed by the topoisomerases.

The RNA primers of the leading strands and Okazaki fragments are removed by the repair DNA polymerases simultaneously replacing the ribonucleotides with deoxyribonucleotides. The gaps that exist between the 3'-OH of one leading strand and the 5'-phosphate of another as well as between one Okazaki fragment and another are repaired by DNA ligases thereby, completing the process of replication.

DNA Replication in Bacteria

DNA is replicated by uncoiling of the helix, strand separation by breaking of the hydrogen bonds between the complementary strands, and synthesis of two new strands by complementary base pairing. Replication begins at a specific site in the DNA. This specific site is known as origin of replication.

DNA replication is bidirectional from the origin of replication. To begin DNA replication, unwinding enzymes called DNA helicases cause the two parent DNA strands to unwind and separate from one another at the origin of replication to form two "Y"-shaped replication forks. These replication forks are the actual sites of DNA copying. Helix destabilizing proteins bind to the single-stranded regions so the two strands do not rejoin. Topoisimerases enzymes produce breaks in the DNA and then rejoin them in order to relieve the stress in the helical molecule during replication.

As the strands continue to unwind and separate in both directions around the entire DNA molecule, new complementary strands are produced by the hydrogen bonding of free DNA nucleotides with those on each parent strand. In the presence of DNA polymerases the new nucleotides line up opposite each parent strand by hydrogen bonding, and join these nucleotides by way of phosphodiester bonds. Actually, the nucleotides lining up by complementary base pairing are deoxynucleoside triphosphates, composed of a nitrogenous base, deoxyribose, and three phosphates. As the phosphodiester bond forms between the 5' phosphate group of the new nucleotide and the 3' OH of the last nucleotide in the DNA strand, two of the phosphates are removed providing energy for bonding. In the end, each parent strand serves as a template to synthesize a complementary copy of it self, resulting in the formation of two identical DNA molecules. During the replication process, the DNA molecules may attach to the cytoplasmic membrane and, as the cell elongates, the two DNA molecules are physically separated.

In reality, DNA replication is more complicated than this because of the nature of the DNA polymerases. DNA polymerase enzymes are only able to join the phosphate group at the 5' carbon of a new nucleotide to the hydroxyl (OH) group of the 3' carbon of a nucleotide already in the chain. As a result, DNA can only be synthesized in a 5' to 3' direction while copying a parent strand running in a 3' to 5' direction. Remember as mentioned above, each DNA strand has two ends. The 5' end of the DNA is the one with the terminal phosphate group on the 5' carbon of the deoxyribose; the 3' end is the one with a terminal hydroxyl (OH) group on the deoxyribose of the 3' carbon of the deoxyribose. The two strands are antiparallel, that is they run in opposite directions. Therefore, one parent strand - the one running 3' to 5' and called the leading strand - can be copied directly down its entire length. However, the other parent strand - the one running 5' to 3' and called the lagging strand - must be copied discontinuously in short fragments of around 100-1000 nucleotides each as the DNA unwinds.

The DNA polymerase enzymes cannot begin a new DNA chain from scratch. They can only attach new nucleotides onto 3' OH group of a nucleotide in a preexisting strand. Thus, to start the leading strand synthesis and each DNA fragment of the lagging strand can be proceed in the presence of, an RNA polymerase complex (i.e. primase) is required. The primase, which is capable of joining RNA nucleotides without requiring a preexisting strand of nucleic acid, first adds several comlementary RNA nucleotides opposite the DNA nucleotides on the parent strand. This forms what is called an RNA primer.

DNA polymerase III then replaces the primase and is able to add DNA nucleotides to the RNA primer. Later, DNA polymerase II digests away the RNA primer and replaces the RNA nucleotides of the primer with the proper DNA nucleotides to fill the gap. Finally, the DNA fragments themselves are hooked together by the enzyme DNA ligase. Yet even with this complicated procedure, a 1000 micrometer-long macromolecule of tightly-packed, supercoiled DNA can make an exact copy of itself in only about 10 minutes time under optimum conditions, inserting nucleotides at a rate of about 1000 nucleotides per second!

Some antimicrobials such as inhibit normal nucleic acid replication in bacteria.

1. The fluoroquinolones (ciprofloxacin, enoxacin, fleroxacin, lomefloxacin, norfloxacin, trovafloxacin, etc.) work by inhibiting one or more of a group of enzymes called topoisomerase, enzymes needed for bacterial nucleic acid synthesis. For example, DNA gyrase (topoisomerase II), breaks and rejoins the strands of bacterial DNA to relieve the stress of the unwinding of DNA that occurs during DNA replication and transcription.
2. The sulfonamides and trimethoprim (co-trimoxazole - a combination of sulfamethoxazole and trimethoprim, sulfanilamide, etc.) block enzymes in the bacteria pathway required for the synthesis of tetrahydrofolic acid, a cofactor needed for bacteria to make the nucleotide bases thymine, guanine, uracil, and adenine. This is done through a process called competitive antagonism whereby a drug chemically resembles a substrate in a metabolic pathway. Because of their similarity, either the drug or the substrate can bind to the substrate's enzyme. While the enzyme is bound to the drug, it is unable to bind to its natural substrate and that blocks that step in the metabolic pathway. Sulfonamides such as sulfamethoxazole tie up the first enzyme in the pathway, the conversion of para-aminobenzoic acid to dihydropteroic acid. Trimethoprim binds to the third enzyme in the pathway, an enzyme that is responsible for converting dihydrofolic acid to tetrahydrofolic acid. Without the tetrahydrofolic acid, the bacteria can not synthesize DNA or RNA.
3. Metronidazole is a drug that is activated by the microbial proteins flavodoxin and feredoxin found in microaerophilc and anaerobic bacteria and certain protozoans. Once activated, the metronidazole puts nicks in the microbial DNA strands.

DNA Replication in Eukaryotes

As in prokaryotes, the linear chromosomes of eukaryotes replicate by strand separation and complementary base pairing of free deoxyribonucleotides with those on each parent DNA strand and. As with prokaryotes, DNA replication in eukaryotic cells is bidirectional. However, unlike the circular DNA in prokaryotic cells that usually has a single origin of replication, the linear DNA of a eukaryotic cell contains multiple origins of replication..

As we know that under prokaryotic DNA replication, DNA can only be synthesized in a 5' to 3' direction and all DNA polymerase requires a primer. To solve this problem, the ends of the linear eukaryotic DNA strands, called telomeres, have short, repetitive, noncoding DNA base sequences. A unique enzyme called telomerase binds to the telomeric DNA at the 3' end. The telomerase contains a small RNA template as a cofactor which is copied by DNA nucleotides to extend the 3' end. Once the extension is long enough, primase can assemble a short RNA primer on the lagging strand and DNA replication can proceed in a manner similar to the lagging strand of prokaryotic DNA.

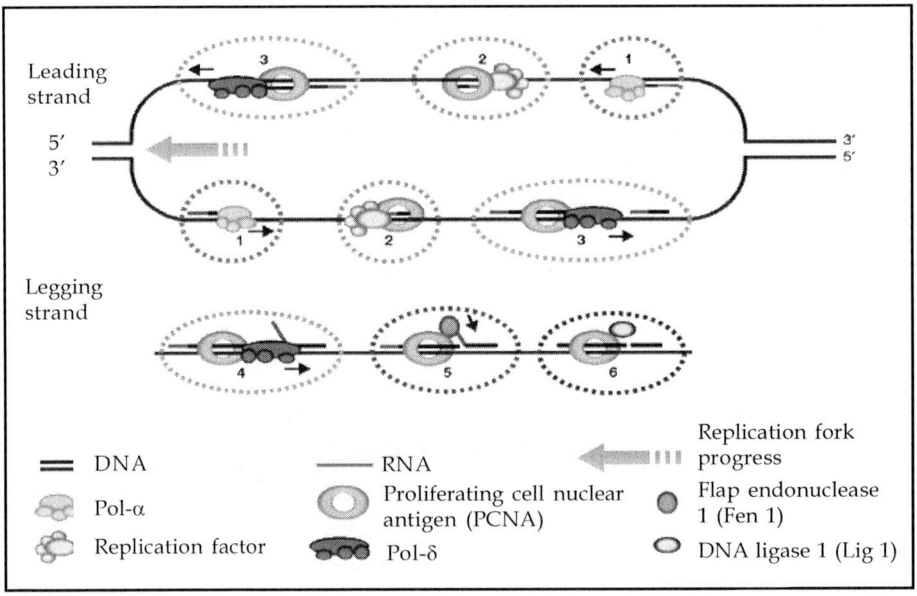

Fig. 5.2. Eukaryotic DNA replication.

Thus the above figure can be summarize as follow:

1. *Step 1:* primer synthesis by DNA polymerase-a (Pol-a);
2. *Step 2:* replication factor C (RFC) displacement of DNA polymerase and recruitment of proliferating cell nuclear antigen (PCNA);

3. *Step 3:* elongation by the newly recruited DNA polymerase-a holoenzyme (Pol-a);
4. *Step 4:* strand displacement by Pol-a;
5. *Step 5:* cutting of the 5' displaced flap by flap endonuclease 1 (Fen1);
6. *Step 6:* sealing by DNA ligase I (LigI).

DNA Polymerase Activities

The main enzymatic activity of DNA polymerases is the 5' → 3' synthetic activity. However, DNA polymerases possess two additional activities of importance for both replication and repair (1). A 5' → 3' exonuclease function and (2). A 3' → 5' exonuclease function. The 5' → 3' exonuclease activity allows the removal of ribonucleotides of the RNA primer, utilized to initiate DNA synthesis, along with their simultaneous replacement with deoxyribonucleotides by the 5' → 3' polymerase activity. The 5' → 3' exonuclease activity is also utilized during the repair of damaged DNA. The 3' → 5' exonuclease function is utilized during replication to allow DNA polymerase to remove mismatched bases. It is possible for DNA polymerases to incorporate an incorrect base during replication. These mismatched bases are recognized by the polymerase immediately due to the lack of Watson-Crick base-pairing. The mismatched base is then removed by the 3' → 5' exonuclease activity and the correct base inserted prior to progression of replication.

Chemotherapies Targeting Replication

The class of compounds that have been used the longest as anticancer drugs are the alkylating agents. The major alkylating agents are derived from nitrogenous mustards that were originally developed for use by the military. Commonly used alkylating agents include cyclophosphamide, ifosphamide, decarbazine, chlorambucil and procarbazine.

Alkylating agents function by reacting with and disrupting the structure of DNA. Some agents react with alkyl groups in DNA resulting in fragmentation of the DNA as a consequence of the action of DNA repair enzymes. Some agents catalzye the cross-linking of bases in the DNA which prevents the separation of the two strands during DNA replication. Some agents induce mispairing of nucleotides resulting in permanent mutations in the DNA. Alkylating agents act upon DNA at all stages of the cell cycle, thus they are potent anticancer drugs. However, because of their potency, prolonged use of alkylating agents can lead to secondary cancers, particularly leukemias.

Several classes of anticancer drugs function through interference with the actions of the topoisomerases. Two of these classes are the anthracyclines and the camptothecins.

The anthracyclins were originally isolated from the fungus, *Streptomyces*. Doxorubicin (Adriamycin, Doxil,) and daunorubicin (Cerubidine, DaunoXome, Daunomycin) have similar modes of action, although doxorubicin is the more potent of the two and is used in the treatment breast cancers, lymphomas and sarcomas. The anthracyclins inhibit the actions of topoisomerase II whose function is to introduce double-strand breaks in DNA during the process of replication as a means to relive torsional stresses. Anthracyclines also function by inducing the formation of oxygen free radicals that cause DNA strand breaks resulting in inhibition of replication.

Another plant-derived anticancer compound that functions through inhibition of topoisomerase II is etoposide (VP-16, Vepesid, Etophos, Eposin). Etoposide is isolated from the mandrake plant and is in a class of compounds referred to as epipodophyllotoxins.

The camptothecins were originally found in the bark of the *Camptotheca accuminata* tree and include irinotecan (Campto, Camptosar) and topotecan (Hycamtin). Camptothecins inhibit the action of topoisomerase I, an enzyme that induces single-strand breaks in DNA during replication.

Anticancer compounds that are extracted from the periwinkle plant, *Vinca rosea*, are called the vinca alkaloids. These compounds bind to tubulin monomers leading to the disruption of the microtubules of the mitotic spindle fibers that are necessary for cell division during mitosis. There are four major vinca alkaloids that are currently used as chemotherapeutics, vincristine (Oncovin, Vincrex), vinblastine (Velbe, Velban, Velsar), vinorelbin (VIN, Navelbine), and vindesine (Eldisine).

The taxanes are another class of plant-derived compounds that act via interference with microtubule function. These compounds are isolated from the Pacific yew tree, *Taxus brevifolia* and include paclitaxel (Taxol) and docetaxel (Taxotere). The taxanes function by hyperstabilizing microtubules which prevents cell division (cytokinesis). These compounds are used to treat a wide range of cancers including head and neck, lung, ovarian, breast, bladder, and prostate cancers. Taxol has proven highly effective in the treatment of certain forms of breast cancer.

In order for DNA replication to proceed, proliferating cells require a pool of nucleotides. The class of anticancer drugs that has been developed to interfere with aspects of nucleotide metabolism is known as the antimetabolites. There are two major types of antimetabolites used in the treatment of a broad range of cancers: componds that inhibit thymidylate synthase and compounds that inhibit dihydrofolate reductase (DHFR). Both of these enzymes are involved in thymidine nucleotide biosynthesis. Drugs that inhibit thymidylate synthase include 5-fluorouracil (5-FU, Adrucil, Efudex) and 5-fluorodeoxyuridine. Those that inhibit DHFR arc analogs of the vitamin folic acid and include methotrexate (Trexall, Rheumatrex) and trimethoprim (Proloprim, Trimpex).

5.2 Transcription

Transcription is the mechanism by which a template strand of DNA is utilized by specific RNA polymerases to generate one of the three different types of RNA, Messenger RNAs (mRNAs), transfer RNAs (tRNAs) and ribosomal RNAs (rRNAs).

1. *Messenger RNAs (mRNAs):* This type of RNAs are the genetic coding templates used by the translational machinery to determine the order of amino acids incorporated into an elongating polypeptide in the process of translation.
2. *Transfer RNAs (tRNAs):* This type of small RNAs form covalent attachments to individual amino acids and recognize the encoded sequences of the mRNAs to allow correct insertion of amino acids into the elongating polypeptide chain.
3. *Ribosomal RNAs (rRNAs):* This type of RNAs are assembled, together with numerous ribosomal proteins, to form the ribosomes. Ribosomes engage the mRNAs and form a catalytic domain into which the tRNAs enter with their attached amino acids. The proteins of the ribosomes catalyze all of the functions of polypeptide synthesis.

All RNA polymerases are dependent upon a DNA template in order to synthesize RNA. The resultant RNA is, therefore, complimentary to the template strand of the DNA duplex and identical to the non-template strand. The non-template strand is called the coding strand due to its' sequences are identical to those of the mRNA. However, in RNA, U is substituted for T.

Classes of RNA Polymerases

In prokaryotic cells, all 3 RNA classes are synthesized by a single polymerase. In eukaryotic cells there are 3 distinct classes of RNA polymerase, *RNA polymerase (pol) I, II and III*. Each polymerase is responsible for the synthesis of a different class of RNA. The capacity of the various polymerases to synthesize different RNAs was shown with the toxin a-amanitin. At low concentrations of a-amanitin synthesis of mRNAs are affected but not rRNAs nor tRNAs. At high concentrations, both mRNAs and tRNAs are affected. These observations have allowed the identification of which polymerase synthesizes which class of RNAs. RNA pol I is responsible for rRNA synthesis.There are 4 major rRNAs in eukaryotic cells designated by there sedimentation size. The 28*S*, 5*S* 5.8*S* and 18S. From that 28*S*, 5*S* and 5.8*S* RNAs are associated with the large ribosomal subunit while the 18*S* rRNA is associated with the small ribosomal subunit. RNA pol II synthesizes the mRNAs and some of the small nuclear RNAs (snRNAs) involved in RNA splicing. RNA pol III synthesizes the tRNAs, the 5*S* rRNA and some snRNAs.

Mechanism of RNA Polymerases

Synthesis of RNA exhibits several features that are synonymous with DNA replication. RNA synthesis requires accurate and efficient initiation, elongation proceeds in the 5'→3' direction (i.e. the polymerase moves along the template strand of DNA in the 3'→5' direction), RNA synthesis requires distinct and accurate termination. Transcription exhibits several features that are distinct from replication.

1. Transcription initiates, both in prokaryotes and eukaryotes, from many more sites than replication.
2. There are many more molecules of RNA polymerase per cell than DNA polymerase.
3. RNA polymerase proceeds at a rate much slower than DNA polymerase.

4. Finally the fidelity of RNA polymerization is much lower than DNA. This is allowable since the aberrant RNA molecules can simply be turned over and new correct molecules made.

Mechanism of Transcription

Signals are present within the DNA template that act in *cis* to stimulate the initiation of transcription. These sequence elements are termed promoters. Promoter sequences promote the ability of RNA polymerases to recognize the nucleotide at which initiation begins. Additional sequence elements are present within genes that act in *cis* to enhance polymerase activity even further. These sequence elements are termed enhancers. Transcriptional promoter and enhancer elements are important sequences used in the control of gene expression.

E. coli RNA polymerase is composed of 5 distinct polypeptide chains. Association of several of these generates the RNA polymerase holoenzyme. The *sigma* subunit is only transiently associated with the holoenzyme. This subunit is required for accurate initiation of transcription by providing polymerase with the proper cues that a start site has been encountered.

In both prokaryotic and eukaryotic transcription the first incorporated ribonucleotide is a purine and it is incorporated as a triphosphate. In *E. coli* several additional nucleotides are added before the sigma subunit dissociates. Elongation involves the addition of the 5'-phosphate of ribonucleotides to the 3'-OH of the elongating RNA with the concomitant release of pyrophosphate. Nucleotide addition continues until specific termination signals are encountered. Following termination the core polymerase dissociates from the template. The core and sigma subunit can then reassociate forming the holoenzyme again ready to initiate another round of transcription.

In *E. coli* transcriptional termination occurs by both factor-dependent and factor-independent means. Two structural features of all *E. coli* factor-independently terminating genes have been identified. One feature is the presence of 2 symmetrical GC-rich segments that are capable of forming a stem-loop structure in the RNA and the second is a downstream A rich sequence in the template. The formation of the stem-loop in the RNA destabilizes the association between polymerase and the DNA template. This is further destabilized by the weaker nature of the AU base pairs that are formed, between the template and the RNA, following the stem-loop. This leads to dissociation of polymerase and termination of transcription.

Factor-dependent termination requires the recognition of termination sequences by the termination protein, *rho*. The rho factor recognizes and binds to sequences in the 3' portion of the RNA. This binding destabilizes the polymerase-template interaction leading to dissociation of the polymerase and termination of transcription.

Post-transcriptional Processing of RNAs

When transcription of bacterial rRNAs and tRNAs is completed they are immediately ready for use in translation. No additional processing takes place. Translation of bacterial mRNAs can begin even before transcription is completed due to the lack of the nuclear-cytoplasmic separation that exists in eukaryotes. The ability to initiate translation of prokaryotic RNAs while transcription is still in progress affords a unique opportunity for regulating the transcription of certain genes. An additional feature of bacterial mRNAs is that most are polycistronic. This means that multiple polypeptides can be synthesized from a single primary transcript. This does not occur in eukaryotic mRNAs.

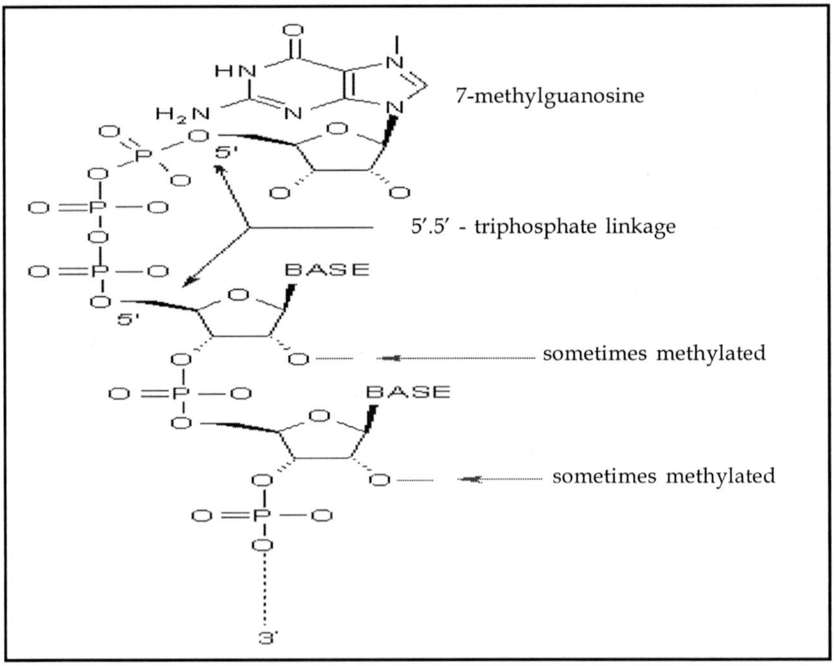

Fig. 5.3. Structure of the 5'-Cap of Eukaryotic mRNAs

In contrast to bacterial transcripts, eukaryotic RNAs undergo significant post-transcriptional processing. All 3 classes of RNA are transcribed from genes that contain introns. The sequences encoded by the intronic DNA must be removed from the primary transcript prior to the RNAs being biologically active. The process of intron removal is called RNA splicing. Additional processing occurs to mRNAs. The 5' end of all eukaryotic mRNAs are capped with a unique 5'→3' linkage to a 7-methylguanosine residue. The capped end of the mRNA is thus, protected from exonucleases and more importantly is recognized by specific proteins of the translational machinery.

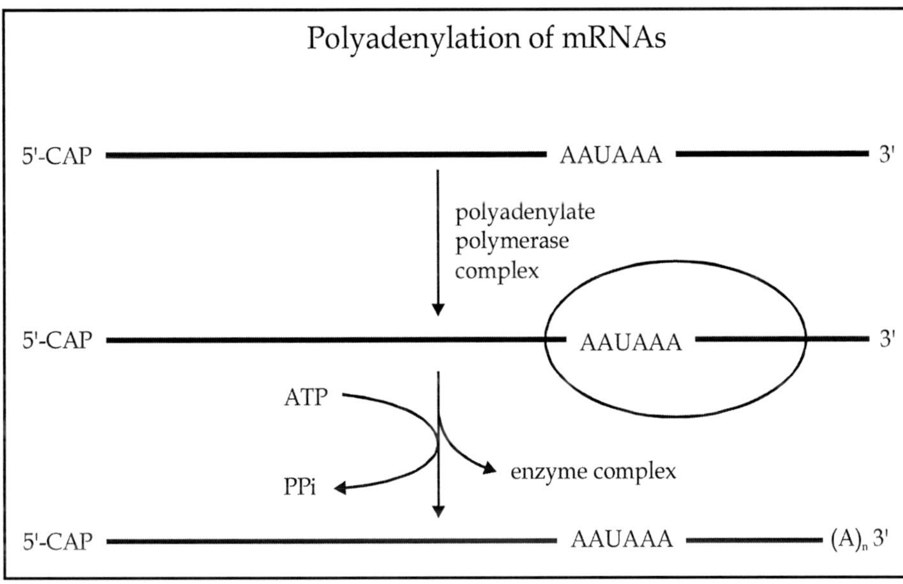

Fig. 5.4 Processes of Polyadenylation

Messenger RNAs also are polyadenylated at the 3' end. A specific sequence, AAUAAA, is recognized by the endonuclease activity of by *polyadenylate polymerase* which cleaves the primary transcript approximately 11 - 30 bases 3' of the sequence element. A stretch of 20-250 A residues is then added to the 3' end by the polyadenylate polymerase activity.

In addition to intron removal in tRNAs, extra nucleotides at both the 5' and 3' ends are cleaved, the sequence 5'-CCA-3' is added to the 3' end of all tRNAs and several nucleotides undergo modification. There have been more than 60 different modified bases identified in tRNAs.

Both prokaryotic and eukaryotic rRNAs are synthesized as long precursors termed preribosomal RNAs. In eukaryotes a 45S preribosomal RNA serves as the precursor for the 18S, 28S and 58S rRNAs.

Splicing of RNAs

There are 4 different classes of introns. The 2 most common are the group I and group II introns. Group I introns are found in nuclear, mitochondrial and chloroplast rRNA genes, group II in mitochondrial and chloroplast mRNA genes. Many of the group I and group II introns are self-splicing, i.e. no additional protein factors are necessary for the intron to be accurately and efficiently spliced out.

Group I Self-Splicing Introns

Group I introns require an external guanosine nucleotide as a cofactor. The 3'-OH of the guanosine nucleotide acts as a nucleophile to attack the 5'-phosphate of the 5' nucleotide of the intron. The resultant 3'-OH at the 3' end of the 5' exon then attacks the 5' nucleotide of the 3' exon releasing the intron and covalently attaching the two exons together. The 3' end of the 5' exon is termed the splice donor site and the 5' end of the 3' exon is termed the splice acceptor site.

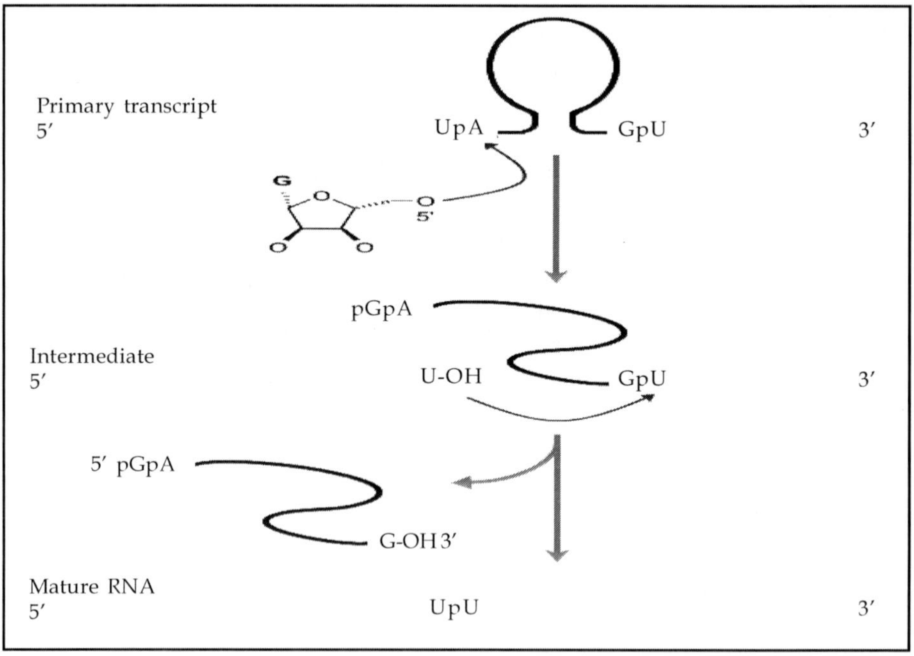

Fig. 5.5 Process of Splicing by Group 1 Introns

Group II Self-Splicing Introns

Here the group II introns are spliced similarly except that instead of an external nucleophile the 2'-OH of an adenine residue within the intron is the nucleophile. This residue attacks the 3' nucleotide of the 5' exon forming an internal loop called a lariat structure. The 3' end of the 5' exon then attacks the 5' end of the 3' exon as in group I splicing releasing the intron and covalently attaching the two exons together.

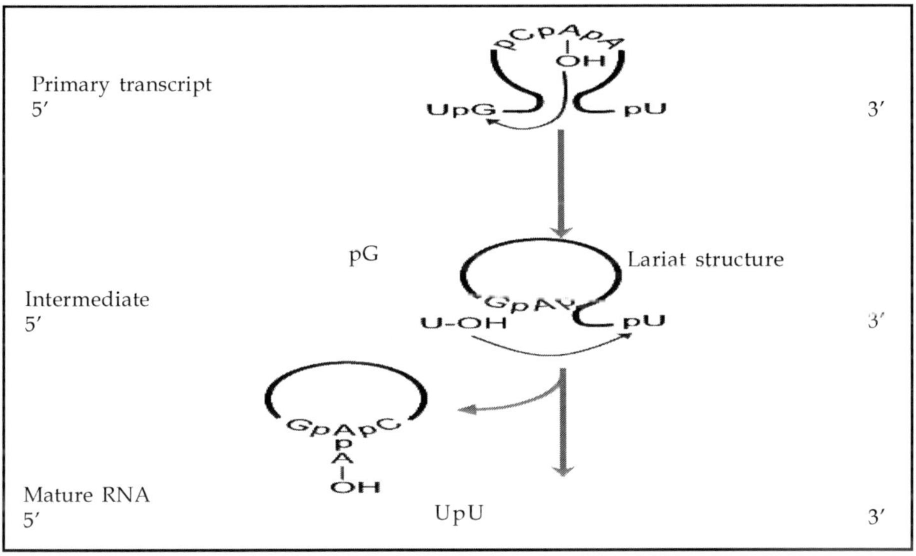

Fig. 5.6. Processes of Splicing by Group 2 Introns

The third class of introns is also the largest class found in nuclear mRNAs. This class of introns undergoes a splicing reaction similar to group II introns in that an internal lariat structure is formed. However, the splicing is catalyzed by specialized RNA-protein complexes called small nuclear ribonucleoprotein particles (snRNPs, pronounced "snurps"). The RNAs found in snRNPs are identified as U1, U2, U4, U5 and U6. The genes encoding these snRNAs are highly conserved in vertebrate and insects and are also found in yeasts and slime molds indicating their importance.

Analysis of a large number of mRNA genes has led to the identification of highly conserved consensus sequences at the 5' and 3' ends of essentially all mRNA introns.

The U1 RNA has sequences that are complimentary to sequences near the 5' end of the intron. The binding of U1 RNA distinguishes the

GU at the 5' end of the intron from other randomly placed GU sequences in mRNAs. The U2 RNA also recognizes sequences in the intron, in this case near the 3' end. The addition of U4, U5 and U6 RNAs forms a complex identified as the spliceosome that then removes the intron and joins the two exons together.

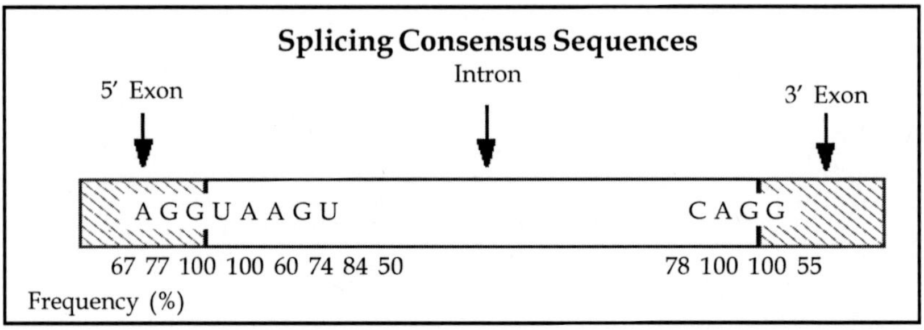

The fourth class of introns are those found in certain tRNAs. These introns are spliced by a specific splicing endonuclease that utilizes the energy of ATP hydrolysis to catalyze intron removal and ligation of the two exons together.

5.3 Translation

Translation is the RNA directed synthesis of polypeptides. This process requires all three classes of RNA. Although the chemistry of peptide bond formation is relatively simple, the processes leading to the ability to form a peptide bond are exceedingly complex. The template for correct addition of individual amino acids is the mRNA, yet both tRNAs and rRNAs are involved in the process. The tRNAs carry activated amino acids into the ribosome which is composed of rRNA and ribosomal proteins. The ribosome is associated with the mRNA ensuring correct access of activated tRNAs and containing the necessary enzymatic activities to catalyze peptide bond formation.

Historical Perspectives

Early genetic experiments demonstrated:

1. The colinearity between the DNA and protein encoded by the DNA. Yanofsky showed that order of observed mutations in the *E. coli* tryptophan synthetase gene was the same as the corresponding amino acid changes in the protein.
2. Crick and Brenner demonstrated, from a large series of double mutants of the bacteriophage T4, that the genetic code is read in

a sequential manner starting from a fixed point in the gene, the code was most likely a triplet and that all 64 possible combinations of the 4 nucleotides code for amino acids, i.e. the code is degenerate since there are only 20 amino acids.

The above mentioned experiments only indicated deductive correlation's regarding the genetic code. The precise dictionary of the genetic code was originally determined by the use of in vitro translation systems derived from *E. coli* cells. Synthetic polyribonucleotides were added to these translation system along with all twenty amino acids. One amino acid at a time was radiolabeled. The first demonstration of the dictionary of the genetic code was with the use of poly(U). This synthetic polyribonucleotide encoded the amino acid phenylalanine, i.e. the resulting polypeptide was poly(F).

The utilization of a variety of repeating di- tri- and tetra polyribonucleotides established the entire genetic code. These results of these experiments confirmed that some amino acids are encoded for by more than one triplet codon, hence the degeneracy of the genetic code. These experiments also established the identity of translational termination codons.

An additional important point to come from these early experiments was that the 5' end of the RNA corresponded to the amino terminus of the polypeptide. This was important since previous labeling experiments had demonstrated that the N-terminus is the beginning of the elongating polypeptide. Therefore, *in vitro* translation experiments established that the RNA is read in the 5' to 3' direction.

Crick first postulated that translation of the genetic code would be carried out through mediation of adapter molecules. Each adapter was postulated to carry a specific amino acid and to recognize the corresponding codon. He suggested that the adapters contain RNA because codon recognition could then occur by complimentarity to the sequences of the codons in the mRNA.

During the course of *in vitro* protein synthesis and labeling experiments it was shown that the amino acids became transiently bound to a low molecular weight mass fraction of RNA. This fraction of RNAs have been termed transfer RNAs (tRNAs) since they transfer amino acids to the elongating polypeptide. These results indicate that accurate translation requires two equally important recognition steps:

1. The correct choice of amino acid needs to be made for attachment to the correspondingly correct tRNA.
2. Selection of the correct amino acid-charged tRNA by the mRNA. This process is facilitated by the ribosomes which we will discuss below.

Summary of Experiments to Determine the Genetic Code

1. The genetic code is read in a sequential manner starting near the 5' end of the mRNA. This means that translation proceeds along the mRNA in the 5' $\rightarrow$ 3' direction which corresponds to the N-terminal to C-terminal direction of the amino acid sequences within proteins.
2. The code is composed of a triplet of nucleotides.
3. That all 64 possible combinations of the 4 nucleotides code for amino acids, i.e. the code is degenerate since there are only 20 amino acids.

The precise dictionary of the genetic code was determined with the use of *in vitro* translation systems and polyribonucleotides. The results of these experiments confirmed that some amino acids are encoded by more than one triplet codon, hence the degeneracy of the genetic code. These experiments also established the identity of translational termination codons.

The Genetic Code

	U	C	A	G
U	UUU Phe UUC Phe UUA Leu UUG Leu	UCU Ser UCC Ser UCA Ser UCG Ser	UAU Tyr UAC Tyr UAA End UAG End	UGU Cys UGC Cys UGA End UGG Trp
C	CUU Leu CUC Leu CUA Leu CUG Leu	CCU Pro CCC Pro CCA Pro CCG Pro	CAU His CAC His CAA Gln CAG Gln	CGU Arg CGC Arg CGA Arg CGG Arg
A	AUU Ile AUC Ile AUA Ile AUG Met	ACU Thr ACC Thr ACA Thr ACG Thr	AAU Asn AAC Asn AAA Lys AAG Lys	AGU Ser AGC Ser AGA Arg AGG Arg
G	GUU Val GUC Val GUA Val GUG Val	GCU Ala GCC Ala GCA Ala GCG Ala	GAU Asp GAC Asp GAA Glu GAG Glu	GGU Gly GGC Gly GGA Gly GGG Gly

Activation of amino acids

Activation of amino acids is carried out by a two step process catalyzed by *aminoacyl-tRNA synthetases*. Each tRNA, and the amino acid it carries, are recognized by individual aminoacyl-tRNA synthetases. This means there exists at least 20 different aminoacyl-tRNA synthetases, there are actually at least 21 since the initiator met-tRNA of both prokaryotes and eukaryotes is distinct from the normal met-tRNA.

Activation of amino acids requires energy in the form of ATP and occurs in a two step reaction catalyzed by the aminoacyl-tRNA synthetases. First the enzyme attaches the amino acid to the α-phosphate of ATP with the concomitant release of pyrophosphate. This is termed an aminoacyl-adenylate intermediate. In the second step the enzyme catalyzes transfer of the amino acid to either the 2'- or 3'-OH of the ribose portion of the 3'-terminal A residue of the tRNA generating the activated aminoacyl-tRNA. Although these reaction are freely reversible, the forward reaction is favored by the coupled hydrolysis of PPi.

Accurate recognition of the correct amino acid as well as the correct tRNA is different for each aminoacyl-tRNA synthetase. Since the different amino acids have different R groups, the enzyme for each amino acid has a different binding pocket for its specific amino acid. It is not the anticodon that determines the tRNA utilized by the synthetases. Although the exact mechanism is not known for all synthetases, it is likely to be a combination of the presence of specific modified bases and the secondary structure of the tRNA that is correctly recognized by the synthetases.

It is absolutely necessary that the discrimination of correct amino acid and correct tRNA be made by a given synthetase prior to release of the aminoacyl-tRNA from the enzyme. Once the product is released there is no further way to proof-read whether a given tRNA is coupled to its corresponding tRNA. Erroneous coupling would lead to the wrong amino acid being incorporated into the polypeptide since the discrimination of amino acid during protein synthesis comes from the recognition of the anticodon of a tRNA by the codon of the mRNA and not by recognition of the amino acid. This was demonstrated by reductive desulfuration of cys-$tRNA^{cys}$ with Raney nickel generating ala-$tRNA^{cys}$. Alanine was then incorporated into an elongating polypeptide where cysteine should have been.

The Wobble Hypothesis

As discussed above, 3 of the possible 64 triplet codons are recognized as translational termination codons. The remaining 61 codons might be considered as being recognized by individual tRNAs. Most cells contain isoaccepting tRNAs, different tRNAs that are specific for the same amino acid, however, many tRNAs bind to two or three codons specifying their cognate amino acids. As an example yeast tRNAphe has the anticodon 5'-GmAA-3' and can recognize the codons 5'-UUC-3' and 5'-UUU-3'. It is, therefore, possible for non-Watson-Crick base pairing to occur at the third codon position, i.e. the 3' nucleotide of the mRNA codon and the 5' nucleotide of the tRNA anticodon. This has phenomenon been termed the wobble hypothesis.

Wobble Positions in Anticodon and Codon Interactions

C	A	G	U	I
G	U	C	A	C
		U	G	U
				A

Wobble Positions in Codon and Anticodon Interactions

C	A	G	U
G	U	C	A
I	I	U	G
			I

Above diagram showing the various modified nucleotides of tRNAs that are found in the wobble position in the anticodon. The top half shows the wobble nucleotides of the anticodon is dark and the various nucleotides (light) of the wobble position of the codon that can be found in non-Watson-Crick base pairs. The lower panel illustrates the opposite showing the wobble nucleotides of the codon in dark & the associated wobble nucleotides of the anticodon in light.

Now that we have charged aminoacyl-tRNAs and the mRNAs to convert nucleotide sequences to amino acid sequences we need to bring the two together accurately and efficiently. This is the job of the ribosomes. Ribosomes are composed of proteins and rRNAs. All living organisms need to synthesis proteins and all cells of an organism need to synthesize proteins, therefore, it is not hard to imagine that ribosomes are a major constituent of all cells of all organisms. The make up of the ribosomes, both rRNA and associated proteins are slightly different between prokaryotes and eukaryotes.

Order of Events in Translation

The ability to begin to identify the roles of the various ribosomal proteins in the processes of ribosome assembly and translation was aided by the discovery that the ribosomal subunits will self assemble in vitro from their constituent parts.

Following assembly of both the small and large subunits onto the mRNA, and given the presence of charged tRNAs, protein synthesis can take place. To reiterate the process of protein synthesis:

1. Synthesis proceeds from the N-terminus to the C-terminus of the protein.
2. The ribosomes "read" the mRNA in the 5' to 3' direction.
3. Active translation occurs on polyribosomes (also termed polysomes). This means that more than one ribosome can be bound to and translate a given mRNA at any one time.
4. Chain elongation occurs by sequential addition of amino acids to the C-terminal end of the ribosome bound polypeptide.

Translation proceeds in an ordered process. First accurate and efficient initiation occurs, then chain elongation and finally accurate and efficient termination must occur. All three of these processes require specific proteins, some of which are ribosome associated and some of which are separate from the ribosome, but may be temporarily associated with it.

Initiation

Initiation of translation in both prokaryotes and eukaryotes requires a specific initiator tRNA, $tRNA^{met}_{i}$, that is used to incorporate the initial methionine residue into all proteins. In *E. coli* a specific version of $tRNA^{met}_{i}$ is required to initiate translation, [$tRNA^{fmet}_{i}$] . The methionine attached to this initiator tRNA is formylated. Formylation requires N^{10}-formy-THF and is carried out after the methionine is attached to the tRNA. The fmet-$tRNA^{fmet}_{i}$ still recognizes the same codon, AUG, as regular $tRNA^{met}$. Although $tRNA^{met}_{i}$ is specific for initiation in eukaryotes it is not a formylated $tRNA^{met}$.

The initiation of translation requires recognition of an AUG codon. In the polycistronic prokaryotic RNAs this AUG codon is located adjacent to a Shine-Delgarno element in the mRNA. The Shine-Delgarno element is recognized by complimentary sequences in the small subunit rRNA (16*S* in *E. coli*). In eukaryotes initiator AUGs are generally, but not always, the first encountered by the ribosome. A specific sequence context surrounding the initiator AUG aids ribosomal discrimination. This context is $^{A}/_{G}CC^{A}/_{G}CCAUG^{A}/_{G}$ in most mRNAs.

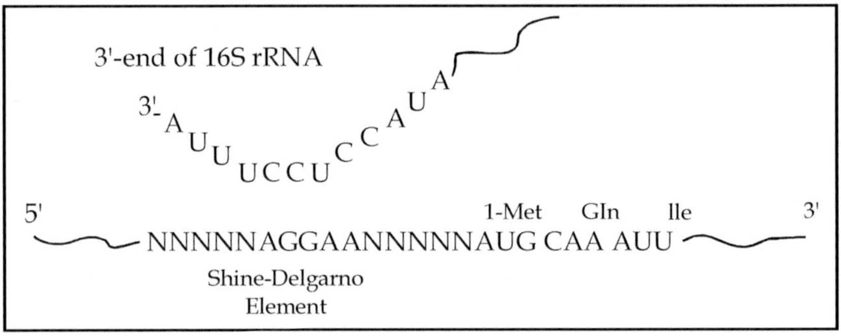

The Shine-Delgarno element is found at the 5' side of each initiator AUG codon in prokaryotic polycistronic mRNAs. This element is complementary to sequences present near the 3'-end of the 16*S* rRNA of the prokaryotic ribosome.

Eukaryotic Initiation Factors and Their Functions

The specific non-ribosomally associated proteins required for accurate translational initiation are termed initiation factors. In *E. coli* they are IFs in eukaryotes they are eIFs. Numerous eIFs have been identified:

Initiation Factor	Activity
eIF-1	repositioning of met-tRNA to facilitate mRNA binding
eIF-2	ternary complex formation
eIF-2A	AUG-dependent met-tRNAmet(i) binding to 40S ribosome
eIF-2B (also called GEF,guanine nucleotide exchange factor)	GTP/GDP exchange during eIF-2 recycling
eIF-3 composed of ~10 subunits	ribosome subunit antiassociation, binding to 40S subunit

Contd....

Initiation factor complex often referred to as eIF-4F composed of 3 primary subunits: eIF-4E, eIF-4A, eIF-4G and at least 2 additional factors: PABP, Mnk1 (or Mnk2)	mRNA binding to 40S subunit, ATPase-dependent RNA helicase activity, interaction between polyA tail and cap structure
PABP: polyA-binding protein	binds to the polyA tail of mRNAs and provides a link to eIF-4G
Mnk1 and Mnk2 eIF-4E kinases	phosphorylate eIF-4E increasing association with cap structure
eIF-4A	ATPase-dependent RNA helicase
eIF-4E	5' cap recognition
4E-BP (also called PHAS) 3 known forms	when de-phosphorylated 4E-BP binds eIF-4E and represses its' activity, phosphorylation of 4E-BP occurs in response many growth stimuli leading to release of eIF-4E and increased translational initiation
eIF-4G	acts as a scaffold for the assembly of eIF-4E and -4A in the eIF-4F complex, interaction with PABP allows 5'-end and 3'-ends of mRNAs to interact
eIF-4B	stimulates helicase, binds simultaneously with eIF-4F
eIF-5	release of eIF-2 and eIF-3, ribosome-dependent GTPase
eIF-6	ribosome subunit antiassociation

Specific Steps in Translational Initiation

Initiation of translation requires 4 specific steps, which are as follows.

1. A ribosome must dissociate into its' 40*S* and 60*S* subunits.
2. A ternary complex termed the preinitiation complex is formed consisting of the initiator, GTP, eIF-2 and the 40*S* subunit.
3. The mRNA is bound to the preinitiation complex.
4. The 60*S* subunit associates with the preinitiation complex to form the 80*S* initiation complex.

The initiation factors eIF-1 and eIF-3 bind to the 40S ribosomal subunit favoring antiassociation to the 60S subunit. The prevention of subunit reassociation allows the preinitiation complex to form. The first step in the formation of the preinitiation complex is the binding of GTP to eIF-2 to form a binary complex. eIF-2 is composed of three subunits α, α and β. The binary complex then binds to the activated initiator tRNA, met-tRNAmet forming a ternary complex that then binds to the 40S subunit forming the 43S preinitiation complex. The preinitiation complex is stabilized by the earlier association of eIF-3 and eIF-1 to the 40S subunit. The cap structure of eukaryotic mRNAs is bound by specific eIFs prior to association with the preinitiation complex. Cap binding is accomplished by the initiation factor eIF-4F. This factor is actually a complex of 3 proteins; eIF-4E, A and G. The protein eIF-4E is a 24 kDa protein which physically recognizes and binds to the cap structure. eIF-4A is a 46 kDa protein which binds and hydrolyzes ATP and exhibits RNA helicase activity. Unwinding of mRNA secondary structure is necessary to allow access of the ribosomal subunits. eIF-4G aids in binding of the mRNA to the 43S preinitiation complex.

Once the mRNA is properly aligned onto the preinitiation complex and the initiator met-tRNAmet is bound to the initiator AUG codon (a process facilitated by eIF-1) the 60S subunit associates with the complex. The association of the 60S subunit requires the activity of eIF-5 which has first bound to the preinitiation complex. The energy needed to stimulate the formation of the 80S initiation complex comes from the hydrolysis of the GTP bound to eIF-2. The GDP bound form of eIF-2 then binds to eIF-2B which stimulates the exchange of GTP for GDP on eIF-2. When GTP is exchanged eIF-2B dissociates from eIF-2. This is termed the eIF-2 cycle (see diagram below). This cycle is absolutely required in order for eukaryotic translational initiation to occur. The GTP exchange reaction can be affected by phosphorylation of the α-subunit of eIF-2. At this stage the initiator met-tRNAmet is bound to the mRNA within a site of the ribosome termed the P-site, for peptide site. The other site within the ribosome to which incoming charged tRNAs bind is termed the A-site, for amino acid site.

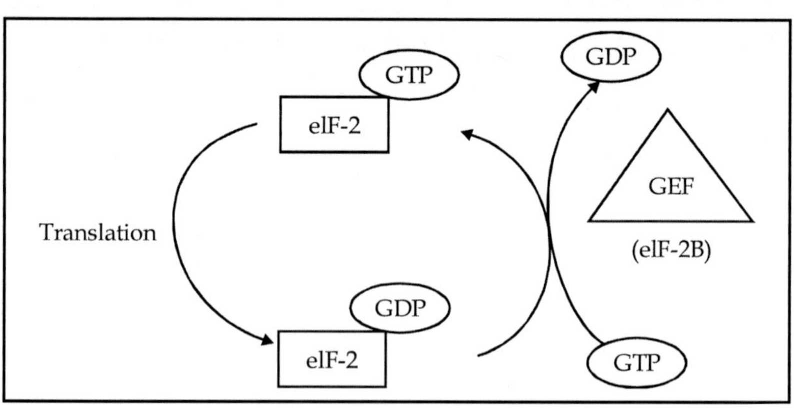

The eIF-2 cycle involves the regeneration of GTP-bound eIF-2 following the hydrolysis of GTP during translational initiation. When the 40*S* preinitiation complex is engaged with the 60*S* ribosome to form the 80*S* initiation complex, the GTP bound to eIF-2 is hydrolyzed providing energy for the process. In order for additional rounds of translational initiation to occur, the GDP bound to eIF-2 must be exchanged for GTP. This is the function of eIF-2B which is also called guanine nucleotide exchange factor (GEF).

Elongation

The process of elongation, like that of initiation requires specific non-ribosomal proteins. In *E. coli* these are EFs and in eukaryotes eEFs. Elongation of polypeptides occurs in a cyclic manner such that at the end of one complete round of amino acid addition the A site will be empty and ready to accept the incoming aminoacyl-tRNA dictated by the next codon of the mRNA. This means that not only does the incoming amino acid need to be attached to the peptide chain but the ribosome must move down the mRNA to the next codon. Each incoming aminoacyl-tRNA is brought to the ribosome by an eEF-1α-GTP complex. When the correct tRNA is deposited into the A site the GTP is hydrolyzed and the eEF-1α-GDP complex dissociates. In order for additional translocation events the GDP must be exchanged for GTP. This is carried out by eEF-1β similarly to the GTP exchange that occurs with eIF-2 catalyzed by eIF-2B. The peptide attached to the tRNA in the P site is transferred to the amino group at the aminoacyl-tRNA in the A site. This reaction is catalyzed by peptidyltransferase. This process is termed transpeptidation. The elongated peptide now resides on a tRNA in the A site. The A site needs to be freed in order to accept the next aminoacyl-tRNA. The process of moving the peptidyl-tRNA from the A site to the P site is termed, translocation. Translocation is catalyzed by eEF-2 coupled to GTP hydrolysis. In the process of translocation the ribosome is moved along the mRNA such that the next codon of the mRNA resides under the A site. Following translocation eEF-2 is released from the ribosome. The cycle can now begin again.

Termination

Like initiation and elongation, translational termination requires specific protein factors identified as releasing factors, RFs in *E. coli* and eRFs in eukaryotes. There are 2 RFs in *E. coli* and one in eukaryotes.

The signals for termination are the same in both prokaryotes and eukaryotes. These signals are termination codons present in the mRNA. There are 3 termination codons, UAG, UAA and UGA.

In *E. coli* the termination codons UAA and UAG are recognized by RF-1, whereas RF-2 recognizes the termination codons UAA and UGA. The eRF binds to the A site of the ribosome in conjunction with GTP. The binding of eRF to the ribosome stimulates the peptidytransferase activity to transfer the peptidyl group to water instead of an aminoacyl-tRNA. The resulting uncharged tRNA left in the P site is expelled with concomitant hydrolysis of GTP. The inactive ribosome then releases its mRNA and the 80S complex dissociates into the 40*S* and 60*S* subunits ready for another round of translation.

Selenoproteins

Selenium is a trace element and is found as a component of several prokaryotic and eukaryotic enzymes that are involved in redox reactions. The selenium in these selenoproteins is incorporated as a unique amino acid, selenocysteine, during translation. A particularly important eukaryotic selenoenzyme is *glutathione peroxidase*. This enzyme is required during the oxidation of glutathione by hydrogen peroxide (H_2O_2) and organic hydroperoxides.

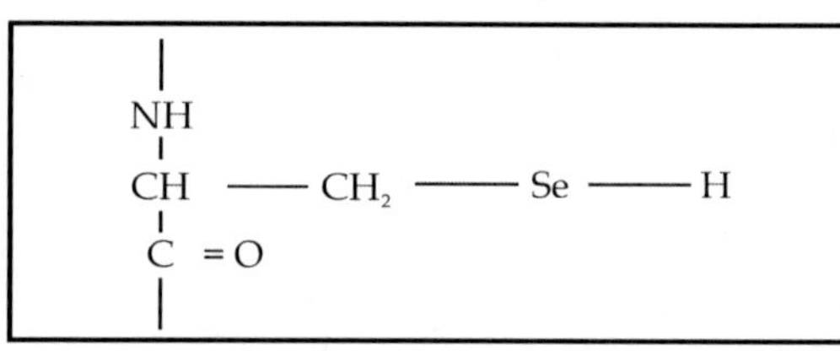

Structure of the selenocysteine residue

Incorporation of selenocysteine by the translational machinery occurs via an interesting and unique mechanism. The tRNA for selenocysteine is charged with serine and then enzymatically selenylated to produce the selenocysteinyl-tRNA. The anticodon of selenocysteinyl-tRNA interacts with a stop codon in the mRNA (UGA) instead of a serine codon. The selenocysteinyl-tRNA has a unique structure that is not recognized by the termination machinery and is brought into the ribosome by a dedicated specific elongation factor. An element in the 3' non-translated region (UTR) of selenoprotein mRNAs determines whether UGA is read as a stop codon or as a selenocysteine codon.

Regulation of eIF-4E Activity

The cellular levels of eIF-4E are the lowest of all eukaryotic initiation factors which makes this factor a prime target for regulation. Indeed, at least 3 distinct mechanisms are known to exist that regulate the level and activity of eIF-4E. These include regulation of the level of transcription of the eIF-4E gene, post-translational modification via phosphorylation and inhibition by interaction with binding proteins. Although the exact mechanisms used to upregulate the transcription of the eIF-4E gene are not yet well understood, it is known that exposure of cells to growth factors as well as activation of T cells leads to increased expression of eIF-4E. The proto-oncogene MYC is believed to play a role in the transcriptional activation of eIF-4E as 2 functional MYC-binding sites have been found in the promoter region of the eIF-4E gene. Of significant note is the finding that cells that are stably over-expressing the MYC gene also have enhanced levels of eIF-4E. Quite strikingly it has been shown that promiscuous elevation in the levels of eIF-4E lead to tumorigenesis placing this translation factor in the category of proto-oncogene. Numerous extracellular stimuli (e.g. insulin, EGF, angiotensis II and gastrin) that exert a portion of their effects at the level of enhanced translation do so by affecting the state of eIF-4E phosphorylation. However, it should be noted that not all signals that lead to increased eIF-4E phosphorylation lead to increased rates of translation. Changes in eIF-4E phosphorylation correlate well with progression through the cell cycle. In resting (G0) cells eIF-4E phosphorylation is low, it increases during G1 and S phase and then declines again in M phase. Phosphorylation of eIF-4E occurs at one major site which is Ser209 (in the human and mouse proteins). The primary signal transduction pathway leading to eIF-4E phosphorylation is that involving the RAS gene. Many growth factors stimulate activation of RAS in response to binding their cognate receptors. Subsequently, RAS activation leads to the phosphorylation and activation of MAP-interacting kinase-1 (Mnk1) which in turn phosphorylates eIF-4E. Although the exact effect of eIF-4E phosphorylation is not clearly defined, it may be necessary to increase affinity of eIF-4E for the mRNA cap structure and for eIF-4G. The principal mechanism utilized in the regulation of eIF-4E activity is through its interaction with a family of binding/repressor proteins termed 4E-BPs (4E binding proteins) which are widely distributed in numerous vertebrate and invertebrate organisms. In mammalian cells 3 related 4E-BPs have been found where 4E-BP1 and 4E-BP2 are also identified as PHAS-I and PHAS-II (PHAS refers to properties of heat and acid stability). Binding of 4E-BPs to eIF-4E does not alter the affinity of eIF-4E for the cap structure but prevents the interaction of eIF-4E with eIF-4G which in turn suppresses the formation of the eIF-4F complex. The ability of 4E-BPs to interact with eIF-4E is controlled via the phosphorylation of specific Ser and Thr residues in 4E-BP. When hypophosphorylated, 4E-BPs bind with high efficiency to eIF-4E but lose their binding capacity when phosphorylated. Numerous growth and signal transduction stimulating effectors lead to phosphorylation of 4E-BPs just as these same responses can lead to phosphorylation of eIF-4E. There are several signal transduction pathways whose activations lead to phosphorylation of 4E-BPs. These include pathways that lead to activation of phosphatidylinositol 3-kinase (PI3K), the Akt Ser/Thr kinase which is also called protein kinase B, PKB (Akt was originally identified as a virally encoded oncogene) and the FKBP12-rapamycin-associated protein/mammalian target of rapamycin (FRAP/mTOR) family of proteins. The mammalian TOR proteins are homologs of the yeast TOR proteins that were identified in a screen for yeast mutants resistant to rapamycin. Rapamycin is an immunosuppressant used primarily in the prevention of tissue rejection following organ transplantation. Rapamycin functions within cells by binding to the immunophilin (an intracellular protein that binds to immunosuppressive drugs, e.g. FK506 and rapamycin) FK506-binding protein 12 (FKBP12). The net effect of phosphorylation of 4E-BPs is that they are released from eIF-4E allowing eIF-4E to actively bind eIF-4G.

Heme Control of Translation

Regulation of initiation in eukaryotes is effected by phosphorylation of a ser(S) residue in the a subunit of eIF-2. Phosphorylated eIF-2 in the absence of eIF-2B is just as active an initiator as non-phosphorylated eIF-2. However, when eIF-2 is phosphorylated the GDP-bound complex is stabilized and exchange for GTP is inhibited. The exchange of GDP for GTP is mediated by eIF-2B (also called guanine nucleotide exchange factor - GEF). When eIF-2 is phosphorylated it binds eIF-2B more tightly thus slowing the rate of exchange. It is this inhibited exchange that affects the rate of initiation.

The phosphorylation of eIF-2 is the result of an activity called heme-controlled inhibitor (HCI) which functions as diagrammed below. HCI is generated in the absence of heme, a mitochondrial product. Removal of phosphate is catalyzed by a specific eIF-2 phosphatase which is unaffected by heme. The presence of HCI was first seen in *in vitro* translation system derived from lysates of reticulocytes. Reticulocytes synthesize almost exclusively hemoglobin at an extremely high rate. In an intact reticulocyte eIF-2 is protected from phosphorylation by a specific 67 kDa protein.

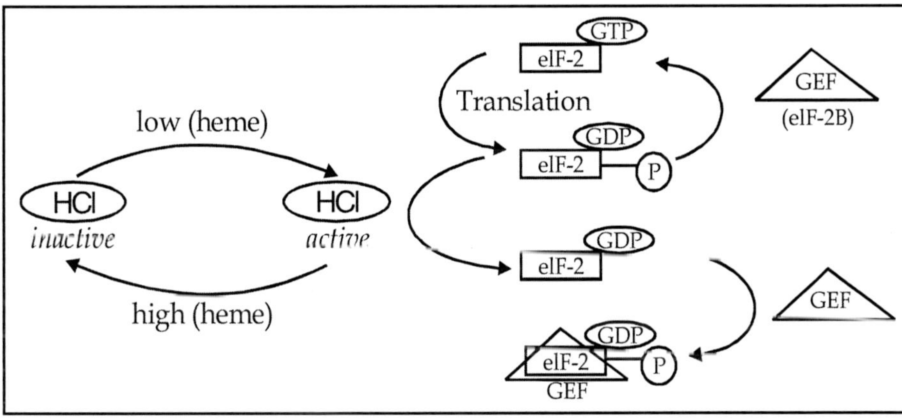

The regulation of translation by heme controlled inhibitor (HCI). Control of translation by heme is clinically important only in erythrocytes. Erythrocytes are enucleate and contain primarily globin mRNA. When the level of heme (required for the synthesis of biologically active hemoglobin) is low it would be inefficient for erythrocytes to synthesize globin protein. As the level of heme falls the activity of HCI increases. HCI is a kinase which phosphorylates eIF-2. When phosphorylated, eIF-2 still hydrolyzes bound GTP to GDP and still interacts with eIF-2B (GEF). However, the rate of eIF-2B-mediated GTP exchange is greatly reduced. This renders eIF-2 incapable of being used to form an new ternary initiation complex and translational initiation is reduced. When the level of heme again rises the activity of HCI is reduced and translational initiation is once again active.

Interferon Control of Translation

Regulation of translation can also be induced in virally infected cells. It would benefit a virally infected cell to turn off protein synthesis to prevent propagation of the viruses. This is accomplished by the induced synthesis of interferons (IFs). There are 3 classes of IFs. The leukocyte or a-IFs, the fibroblast or b-IFs and the lymphocyte or g-IFs. IFs are induced by dsRNAs and themselves induce a specific kinase termed RNA-dependent protein kinase (PKR) that phosphorylates eIF-

2 thereby shutting off translation in a similar manner to that of heme control of translation. Additionally, IFs induce the synthesis of 2′5'-oligoadenylate, pppA(2′p5′A)n, that activates a pre-existing ribonuclease, RNase L. RNase L degrades all classes of mRNAs thereby shutting off translation.

Iron Control of Translation

Regulation of the translation of certain mRNAs occurs through the action of specific RNA-binding proteins. Protein of this class have been identified that bind to sequences in either the 5' non-translated region (5'-UTR) or 3'-UTR. Two particularly interesting and important regulatory schemes related to iron metabolism encompass RNA binding proteins that bind to either the 5'-UTR of one mRNA or the 3'-UTR of another.The transferrin receptor is a protein located in the plasma membrane that binds the protein transferrin. Transferrin is the major iron transport protein in the plasma. When iron levels are low the rate of synthesis of the transferrin receptor mRNA increases so that cells can take up more iron. This regulation occurs through the action of an iron response element binding protein (IRBP) that binds to specific iron response elements (IREs) in the 3'-UTR of the transferrin receptor mRNA. These IREs form hair-pin loop structures that are recognized by IRBP. This IRBP is an iron-deficient form of aconitase, the iron-requiring enzyme of the TCA cycle. When iron levels are low, IRBP is free of iron and can therefore, interact with the IREs in the 3'-UTR of the transferrin receptor mRNA. Transferrin receptor mRNA with IRBP bound is stabilized from degradation. Conversely, when iron levels are high, IRBP binds iron then cannot interact with the IREs in the transferrin receptor mRNA. The effect is an increase in degradation of the transferrin receptor mRNA. A related, but opposite, phenomenon controls the translation of the ferritin mRNA. Ferritin is an iron-binding protein that prevents toxic levels of ionized iron (Fe^{2+}) from building up in cells. The ferritin mRNA has an IRE in its 5'-UTR. As with the transferrin receptor story, when iron levels are high, IRBP cannot bind to the IRE in the 5'-UTR of the ferritin mRNA. This allows the ferritin mRNA to be translated. Conversely, when iron levels are low, the IRBP binds to the IRE in the ferrritin mRNA preventing its translation.

Protein Synthesis Inhibitors

Many of the antibiotics utilized for the treatment of bacterial infections as well as certain toxins function through the inhibition of translation. Inhibition can be effected at all stages of translation from initiation to elongation to termination.

Several Antibiotic and Toxin inhibitors of Translation

Inhibitor	Comments
Chloramphenicol	inhibits prokaryotic peptidyl transferase
Streptomycin	inhibits prokaryotic peptide chain initiation, also induces mRNA misreading
Tetracycline	inhibits prokaryotic aminoacyl-tRNA binding to the ribosome small subunit
Neomycin	similar in activity to streptomycin
Erythromycin	inhibits prokaryotic translocation through the ribosome large subunit
Diptheria toxin	catalyzes ADP-ribosylation of and inactivation of eEF-2
Ricin	found in castor beans, catalyzes cleavage of the eukaryotic large subunit rRNA
Cycloheximide	inhibits eukaryotic *peptidyltransferase*

❑❑❑

6

GENE EXPRESSION

6.1 Prokaryotic gene expression

Control of gene expression is at the level of transcription. If a gene is not transcribed then the gene product and ultimately the phenotype will not be expressed. We are now going to consider two systems of control of gene expression in the *E. coli* cell. Both of these systems are concerned with the production of enzymes involved in cell metabolism, but each exhibits a different type of control.

All of the genes which encode the enzymes necessary for the pathway are found next to each other on the *E. coli* chromosome. One key feature of both systems to be discussed is that a single mRNA is transcribed with multiple translation stop codons. The proteins that can be translated from the mRNA are the enzymes required for a specific pathway. This type of mRNA is called a polycistronic mRNA and is totally unique to prokaryotes.

6.2 The *lac* Operon

Jacob and Monod for which they were awarded the Nobel Prize coined the *lac* operon system. In the *lac* operon system an operon required for the transport and metabolism of lactose in *Escherichia coli* and some other enteric bacteria. It consists of three adjacent structural genes, (1) promoter, (2) terminator, and (3) an operator. The *lac* operon is regulated by several factors including the availability of glucose and of lactose. Gene regulation of the *lac* operon was the first genetic regulatory mechanism to be elucidated and is often used as the canonical example of prokaryotic gene regulation.

The first control system for enzyme production worked out at the molecular level described the control of enzymes that are produced in response to the presence of the sugar lactose in *E. coli* cell. The following is the pathway that leads to the production of glucose and galactose.

$$\text{Lactose} \xrightarrow[\beta\text{-galactosidase}]{} \text{Glucose + Galactose}$$

Several proteins involved in lactose metabolism in the *E. coli* cell. They are as follows:

1. β-galactosidase - which, converts lactose into glucose and galactose
2. β-galactoside permease –which, transports lactose into the cell
3. β-galactoside transacetylase - the function not known

Research with this system was greatly added by the availability of constitutive mutants. A constitutive mutant is one in which the gene product is produced continually, that is there is no control over its expression. In these mutants, the above proteins were produced all the time in comparison to the wild type where the proteins only appeared in the presence of lactose. So in this system, the mutation must be a gene other than those responsible for the structural genes.

All of the genes involved in controlling this pathway are located next to each other on the *E. coli* chromosome. Together they form an operon.

***lac* Operon Gene**	**Gene Function**
I	Gene for repressor protein
P	Promoter
O	Operator
lac Z	Gene for ß-galactosidase
lac Y	Gene for ß-galactoside permease
lac A	Gene for ß-galactoside transacetylase

An operator is a segment of DNA which regulates the activity of the structural genes of the operon that it is linked to, by interacting with a specific repressor or activator. It is a regulatory sequence for shutting a gene down or turning it "on". A promoter is a DNA sequence that enables a gene to be transcribed. The promoter is recognized by RNA polymerase, which then initiates transcription. In RNA synthesis, promoters indicate which genes should be used for messenger RNA creation - and, by extension, control which proteins the cell manufactures.

Control Circuit for the *lac* Operon

I P O || Z | Y | A |

Controlling Region || Structural genes

Fig. 6.1: The genetic structure of the operon.

How does the system work?

In the absence of lactose in the cell, the repressor protein binds to the operator and prevents the read through of RNA polymerase into the three structural genes. With lactose in the cell, lactose binds to the repressor. This causes a structural change in the repressor and it loses its affinity for the operator. Thus RNA polymerase can then bind to the promoter and transcribe the structural genes. In this system lactose acts as an effector molecule.

With the above information, now we can predict the effect that various mutants will have on lac operon gene expression.

Mutant *lac* gene	Mutant Phenotype
I^-	constitutive expression because the operator is never closed
O^-	constitutive expression because the repressor protein can not bind
P^-	no expression of the operon because RNA polymerase cannot bind
lac Z^-	no glucose or galactose production from lactose
lac Y^-	no induction because lactose will not be taken into the cell

Catabolite Repression of the lac Operon

Lactose is not the preferred carbohydrate source for *E. coli*. If lactose and glucose are present, the cell will use all of the glucose before the *lac* operon is turned on. This type of control is termed catabolite repression. To prevent lactose metabolism, a second level of control of gene expression exists. The promoter of the *lac* operon possess two binding sites. One site is the location where RNA polymerase binds and second location is the binding site for a complex between the catabolite activator protein (CAP) and cyclic AMP (cAMP). The binding of the CAP-cAMP complex to the promoter site is required for transcription of the *lac* operon. The presence of this complex is closely associated with the presence of glucose in the cell. As the concentration of glucose increases the amount of cAMP decreases. When the cAMP decreases, the amount of complex decreases. This decrease in the complex inactivates the promoter, and the *lac* operon is turned off. Because the CAP-cAMP complex is needed for transcription, the complex exerts a positive control over the expression of the *lac* operon.

Lac Operon

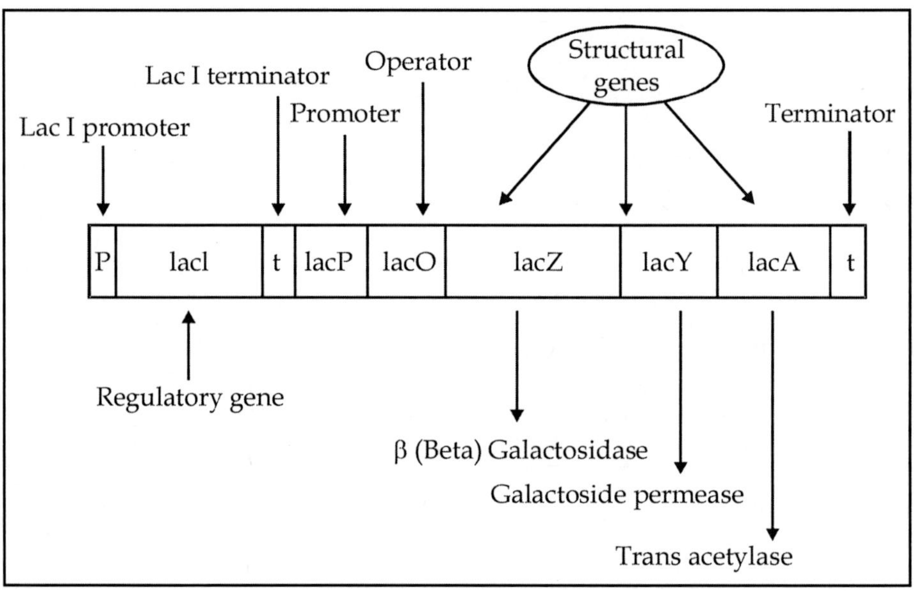

6.3 The *trp* Operon

The *trp* operon of *E. coli* controls the biosynthesis of tryptophan in the cell from the initial precursor chorismic acid. This operon contains genes for the production of five proteins, which are used to produce three enzymes. The products of the *E* and *D* genes form a multimeric protein comprised of two copies of each protein to produce the enzyme anthranilate synthetase. This enzyme catalyzes the first two reactions in the tryptophan pathway. The next enzyme, which is responsible for catalyzing the next two steps in the pathway, is indole glycerolphosphate synthetase and it is the product of the *C* locus. In this reaction the final step is the pathway produces tryptophan from indole-glycerol phosphate and serine. This single step is catalyzed by tryptophan synthetase, which is a multimer of two proteins that are the product of the *B* and *A* genes.

The *trp* operon consists of the repressor, promoter, operator and the structural genes. In this system, though, unlike the *lac* operon, the gene for the repressor is not adjacent to the promoter, but rather is located in another part of the *E. coli* genome. Another difference is that the operator resides entirely within the promoter.

Control Circuit for the *trp* Operon
P/O | L || E | D | C | B | A |
Controlling Region || Structural genes

***trp* Operon Gene**	**Gene Function**
P/O	Promoter; operator sequence is found in the promoter
trp L	Leader sequence; attenuator (A) sequence is found in the leader
trp E	Gene for anthranilate synthetase subunit
trp D	Gene for anthranilate synthetase subunit
trp C	Gene for glycerolphosphate synthetase
trp B	Gene for tryptophan synthetase subunit
trp A	Gene for tryptophan synthetase subunit

Thus the *trp* operon is a repressible system. The primary difference between repressible and inducible systems is the result that occurs when the effector molecule binds to the repressor. With inducible systems, the binding of the effector molecule to the repressor greatly reduces the affinity of the repressor for the operator, the repressor is released and transcription proceeds. The *lac* operon is an example of an inducible system. With repressible systems, the binding of the effector molecule to the repressor greatly increases the affinity of repressor for the operator and the repressor binds and stops transcription. Thus, for the *trp* operon , the addition of tryptophan to the *E. coli* environment shuts off the system because the repressors binds at the operator.

Attenuation of the *trp* Operon

One element of the trp operon is the leader sequence that in immediately 5' of the trpE gene. This sequence about 160 bp is size also controls the expression of the operon through a process called attentuation. This sequence has four domains (1-4). Domains 3 (nucleotides 108-121) of the mRNA can base pair with either domain 2 (nucleotides 74-94) or domain 4 (nucleotides 126-134). If domain 3 pairs with domain 4, a stem and loop structure forms on the mRNA and transcription stops. This structure forms when the level of tryptophan is high in the cell. If domain 3 pairs with domain 2, then the stem and loop structure does not form and transcription continues through the operon, and all of the enzymes required for tryptophan biosynthesis are produce. These events occur when tryptophan is low in the cell. If domain 4 is deleted, the stem and loop structure cannot form and transcription of the remainder of the operon will occur even in the presence of tryptophan. Domain 4 is called the attenuator because its presence is required to reduce mRNA transcription in the presence of high levels of tryptophan. The section of the leader sequence encodes a 14 amino acid peptide containing two tryptophan residues.

Attenuation of the *trp* operon mRNA

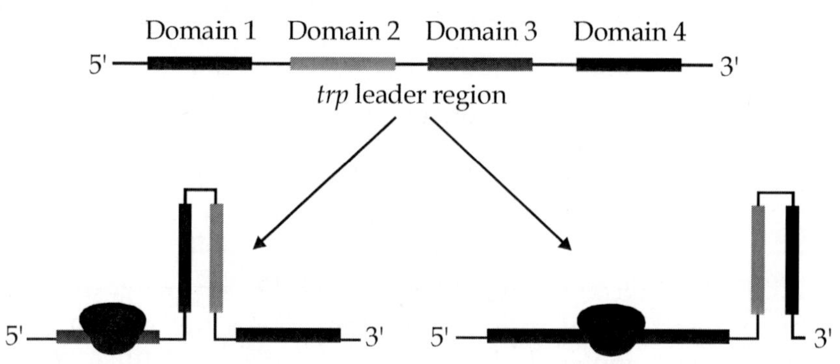

Low Tryptophan Levels
- Slow translation of Domain 1 peptide
- Normal full gene transcription

High Tryptophan Levels
- Fast translation of domain 1 peptide
- Domain 2 blocked by ribosome
- Domain 3-4 pairing occurs
- Attentuation of transcription occurs
- Only 10 % of normal mRNA s made

trp Operon Transcription Under High Levels of Tryptophan

When the cellular levels of tryptophan are high, the levels of the tryptophan tRNA are also high. Immediately after transcription, the mRNA moves quickly through the ribosome complex and the small peptide is translated. Translation is quick because of the high levels of tryptophan tRNA. Because of the quick translation, domain 2 becomes associated with the ribosome complex. Then domain 3 binds with domain 4, and transcription is attenuated because of the stem and loop formation.

trp Operon Transcription Under Low Levels of Tryptophan

Under low cellular levels of tryptophan, the translation of the short peptide on domain 1 is slow. Because of the slow translation, domain 2 does not become associated with the ribosome. Rather domain 2 associates with domain 3. This structure permits the continued transcription of the operon. Then the *trpE-A* genes are translated, and the biosynthesis of tryptophan occurs

❑❑❑

7

TECHNIQUES

7.1 Gel Electrophoresis

Gel electrophoresis is used to characterize one of the most basic properties - molecular mass - of both polynucleotides and polypeptides

- Gel electrophoresis can also be used to determine (a) the *purity* of these samples (b) *heterogeneity*/extent of degradation (c) *subunit composition*.

DNA agarose gels

The electrophoretic migration rate of DNA through agarose gels is dependent upon four main parameters:

1. The molecular size of the DNA. Molecules of linear duplex DNA travel through agarose gels at a rate which is inversely proportional to the log of their molecular weight. $M_r \alpha 1/\log (Mw)$

   Example: Compare molecular mass vs. expected migration rate:

Molecular Mass (Da)	log (Molec. Mass)	1/log (Molec. Mass) *i.e. relative Mr*
100,000	5.0	0.20
50,000	4.7	0.21
10,000	4.0	0.25
5,000	3.7	0.27
1,000	3.0	0.33

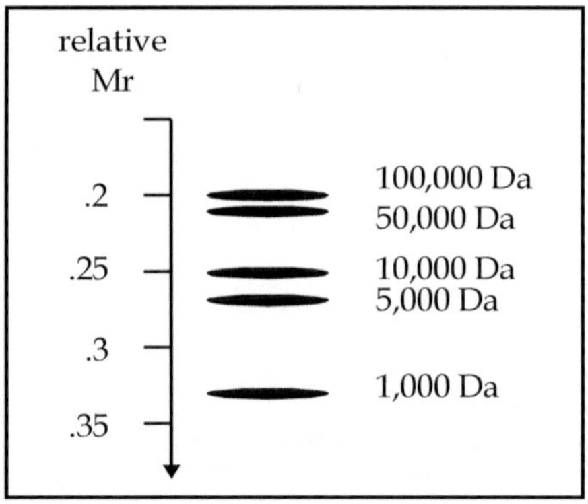

2. The agarose concentration. There is an inverse linear relationship between the logarithm of the electrophoretic mobility and gel concentration.

$\log (M_r) \alpha \; 1/[gel]$

Gel %	**1/Gel %**	**inv log(1/Gel %)** *(i.e. relative Mr)*
2.0	0.50	3.2
1.5	0.67	4.6
1.0	1.00	10.0
0.5	2.00	100.0

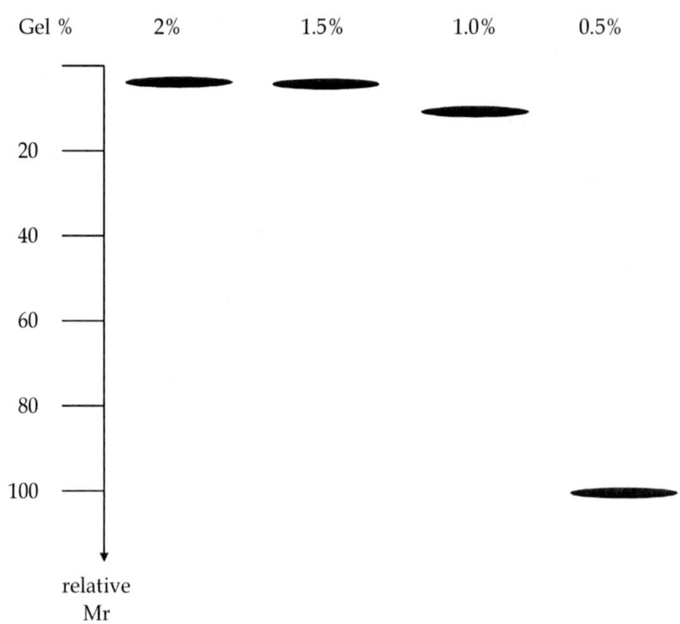

3. The conformation of the DNA. (a). closed circular DNA (form-I) - typically supercoiled (b) nicked circular (form-II), and (c) linear DNA (form-III)

   These different forms of the *same DNA* migrate at different rates through an agarose gel. (a). Almost always the linear form (form-III) migrates at the slowest rate of the three forms (b) Supercoiled DNA (form-I) usually migrates the fastest.

4. The applied voltage. (a) Typical value for running an agarose gel is 5 volts per cm (length of gel). (b) Agarose gels are usually poured and run horizontally

Agarose (%)	**Range of separation of linear DNA (in kilobases)**
0.3	60 - 5
0.6	20 - 1
0.7	10 - 0.8

Thus, the DNA being an acidic molecule and its migrates towards to the positively charged electrode i.e cathode.

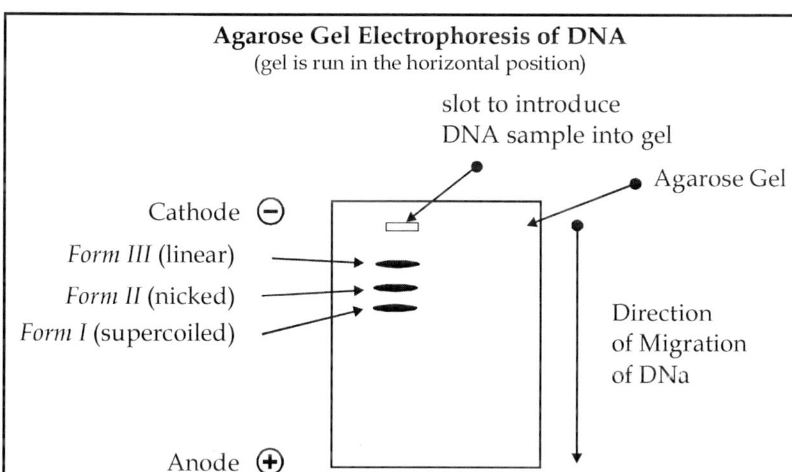

DNA acrylamide gels

Acrylamide gels are useful for separation of small DNA fragments, (a) typically oligonucleotides <100 base pairs. (b) These gels are usually of a low acrylamide concentration (<=6%) and contain the non-ionic denaturing agent Urea (6M).

The denaturing agent prevents secondary structure formation in oligonucleotides and allows a relatively accurate determination of molecular mass.

Stains of DNA

Ethidium

The most convenient method to visualize DNA in gel electrophoresis is staining with the fluorescent dye ethidium bromide. This compound contains a planar group that intercalates between the stacked bases of DNA.

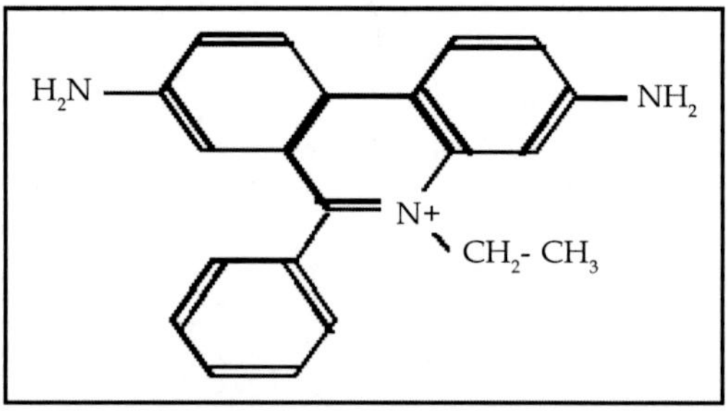

- The orientation and proximity of ethidium with the stacked bases causes the dye to display an increased flourescence compared to free dye (in solution).

  Here the U.V. radiation at 254 nm is absorbed by the DNA and transmitted to the bound dye and the energy is re-emitted at 590 nm in the red-orange region of the spectrum.

- Ethidium bromide is usually prepared as a stock solution of 10 mg/ml in water, stored at room temp and protected from light.
- The dye is usually incorporated into the gel and running buffer, or conversely, the gel is stained after running by soaking in a solution of ethidium bromide (0.5 ug/ml for 30 min).
- The stain is visualized by irradiating with a UV light source (i.e. using a transiluminator) and photgraphing with polaroid film.
- The usual sensitivity of detection is better than 0.1 ug of DNA.

Because ethidium is a DNA intercalating agent, it is a powerful mutagen. Incorporation of ethidium in the DNA of living organisms can cause mutations.

Important Note

1. The intercalation of ethidium bromide causes a reduction in the number of effective base pairs per twist.
2. Thus, the DNA wants to adopt a conformation with fewer base pairs per twist.

   a. The number of twists will increase for the given length of DNA

   b. Writhe will decrease if the Linkage number remains unchanged
3. The net result is that addition of ethidium bromide will introduce negative supercoils into the DNA .
4. Addition of ethidium bromide to a DNA sample prior to electrophoresis can cause positively supercoiled DNA to migrate as relaxed (nicked, or form II) DNA
5. After replication DNA is typically under wound (negative supercoils), thus it will become more under wound with EtBr treatment (migration rate may actually increase due to more compact state).

Flourescence shadowing

- DNA fragments resolved on polyacrylamide gels can also be visualized by the method of UV shadowing.
- In this method the gel is placed on top of a fluorescent material, usually a flourescent TLC silica plate.
- The gel is then illuminated by a UV light source.
- DNA bands in the gel will block transmittance of the UV light to the substrate.
- This will result in a dark area (i.e. non-fluorescing) area on the substrate.
- This method is usually used in gel purification of oligonucleotides.

Molecular Weight Standards

1. DNA gel electrophoresis utilize molecular weight standards to calibrate the size(s) of samples being analyzed.
2. DNA molecular weight standards will consist of a mixture of DNA fragments of known sizes i.e. molecular mass. (i) One convenient DNA molecular weight standard is constructed by partial ligation of a 100 base pair fragment of duplex DNA.

- Partial ligation will result in formation of dimers (200 bp), trimers (300 bp) and so on, as well as some amount of the original (100 bp) fragment
- This produces a DNA "ladder" after gel electrophoresis

(ii) Another DNA fragment standard might be a known DNA sequence, such as the plasmid pBR322, which has been digested with a four-cutter restriction endonuclease (e.g. Alu I).

- This produces a variety of fragments with different sizes
- It is easily reproduced (just cut more pBR322)
- In both DNA and protein gels, molecular weight markers are run in a lane at one end of the gel

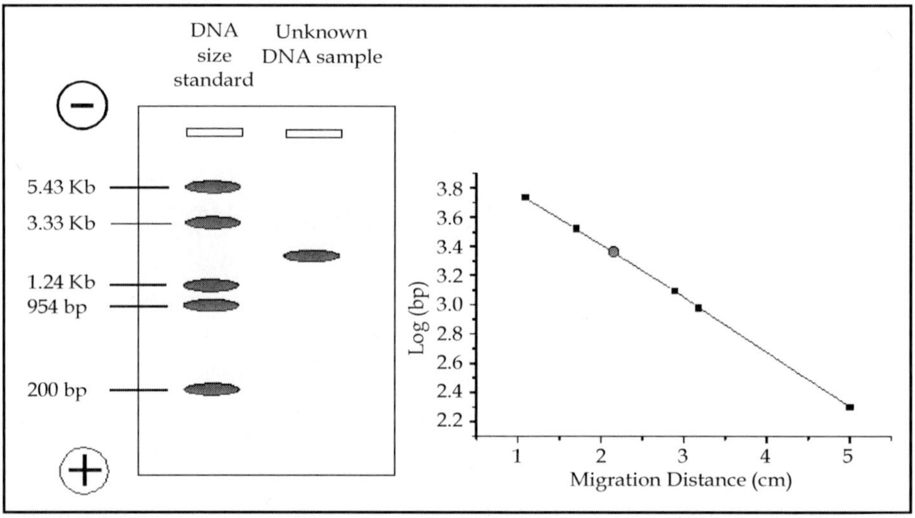

DNA Sequence Analysis

DNA sequencing is most often accomplished using a procedure referred by (1) Sanger sequencing (2) Di-deoxy sequencing and (3) Chain termination sequencing.

- Each of these refers to the same method:
  - The use of di-deoxy base incorporation in a polymerization reaction
  - Leads to termination of primer extension (a method pioneered by Fred Sanger).

There are some basic method involves

1. Annealing a primer 5' to a region of DNA we would like to sequence.
2. The primer is extended in the traditional manner (i.e. with DNA polymerase and the four dNTP's).
3. However, small concentrations of di-deoxy bases are included in the reaction mixture.
   - Usually this is accomplished by having four separate reactions, into which one of the ddNTP's is added.
   - Thus, one tube would contain primer, template, DNA polymerase, the four dNTP's and ddATP, another tube would have the same thing but with ddTTP instead of ddATP, and so on for ddCTP and ddGTP.
4. During the reaction, the normal dNTP's are incorporated into the growing chain.
   - However, occasionally the DNA pol will incorporate a ddbase into the growing primer.
   - When this happens, the primer cannot be extended any further (because the 3' dd base does not have an available 3' hydroxyl group).
   - The resulting DNA fragment begins at the 5' end of the sequencing primer and ends at the site of dd base incorporation.

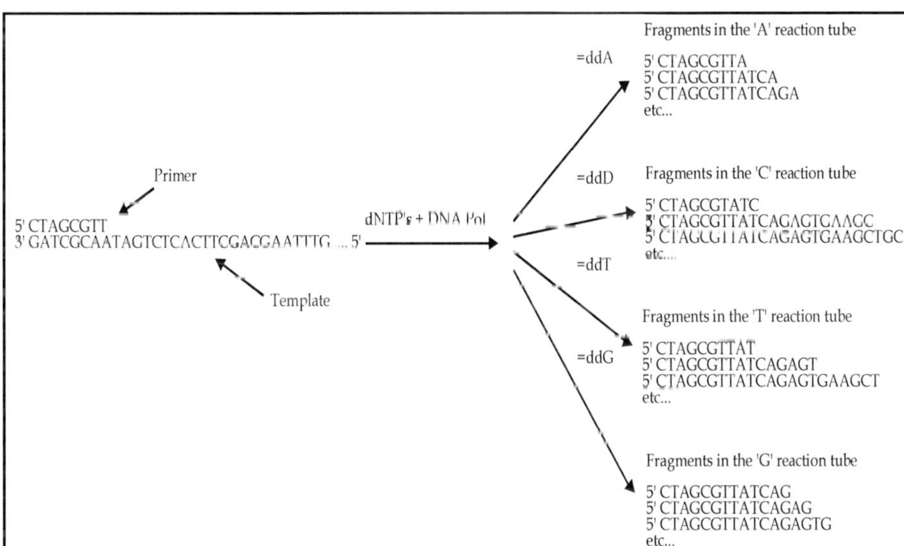

Thus, in the reaction mixture containing the dd base ddATP, there will result an ensemble of fragements of varying lengths, each ending in with the ddA base (i.e. at all positions in the template where there was a comlementary 'T' base).

The mixture containing ddCTP will have a different mix of fragements - they will contain ddC at the 3' ends (at positions in the template where 'G' bases were located).

- If the fragments from the 'A' reaction mix are run on a urea/ acrylamide gel (typically 6%) the fragments will separate according to size.
- Likewise for the 'C', 'G' and 'T' reaction mix fragments.
  - If the four different reaction mixtures are run next to each other the fragment sizes can be directly compared to one another.
  - Note that the shortest expected fragment is the primer itself and the longer the fragment, the further from the primer the extension reaction went before termination.
- Consider a case where the template has a stretch of six 'A' bases in a row.
  - In the 'T' reaction mix we will subsequently get six fragments; each ending in 'T' and differing by one base length.
  - None of the other reaction mixes will contain fragments between these lengths (they will either be longer or shorter) because none of the other reaction mixes will terminate within this region.
  - Thus, if we run the four reaction mixes side by side and look at the fragment patterns we would see the following:

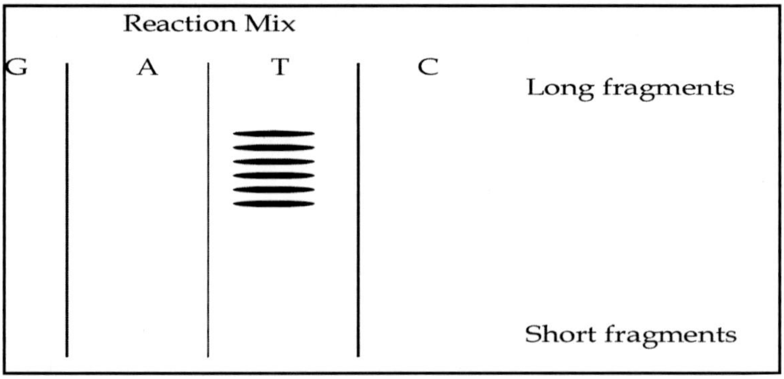

- Now consider a template that contains the sequence 3' GATC 5'.
  - When the primer is extended in the different reaction mixes it can truncate first at the G (incorporating a dd 'C'), then at the A (incorporating a dd 'T') and so on.
  - The fragments run on a gel would thus look like:

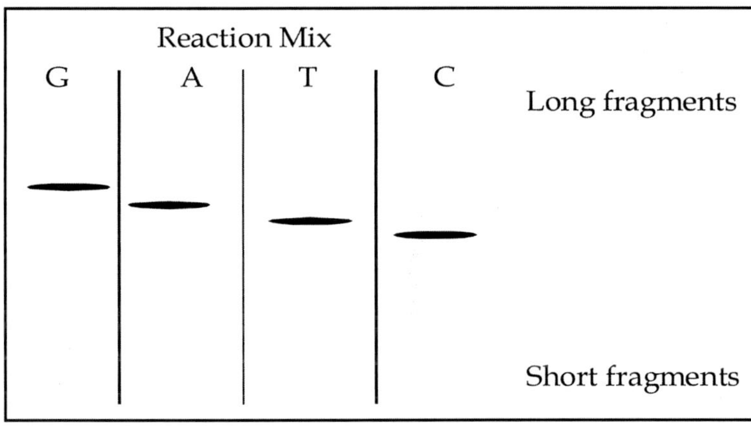

Thus, following the ladder of fragments on the sequencing gel allows to "read" the sequence of the template.

Important Note : In regard to the template when we read from the bottom of the gel to the top we are reading in the 3' to 5' direction and reading the complementary bases to the actual sequence.

Visualization of fragments

If radiolabeled dATP is spiked into the mixtures it will be incorporated like a "normal" dATP base. However, the resulting DNA fragment will be *radiolabled* and thus the acrylamide gel can be exposed to x ray film and the location of the fragments determined.

Recently, automated sequencers have made use of specific dyes which are tagged to the dideoxy bases.

1. These dyes can be "read" by a laser and thus the specific terminating dd base for a particular DNA fragment can be identified.
2. Thus, the fragments can read as they elute, rather than stopping the gel and exposing it.

3. Furthermore, since each dd base can be uniquely identified, all four reactions (ddA, ddC, ddG and ddT chain termination) can be done in a single tube and run in a single lane on a sequencing gel
   - Automated sequencers can thus read further than with manual methods.
   - Since a single lane is used per sample (as compared to four lanes with the radiolabeled method) many more samples can be analyzed and the throughput is greater.
4. The acrylamide gels used for sequence analysis are typically 50 cm to 100 cm long. In manual sequencing the four reaction mixes are loaded and the gel is run for approximately 2 hours then the samples are reloaded on another part of the gel and the gel run is continued. A third set of samples may be loaded after another 2 hours.
   - The gel is stopped after the dye front of last sample loaded has just reached the bottom of the gel.
   - Thus, the short fragments can be visualized in the last load, medium fragements in the second load and the long fragments will be visualized in the first set of reaction mixtures loaded.
   - Manual sequencing can resolve on the order of 400 bases of continuous sequence. Automated sequencers can routinely provide twice this amount of information.
5. Automated sequencers use the same types of glass plates
   - The continuous running of the gel (and dye identification) means that typically 400-700 bases or more can be read.
   - Automatic software will interpret the dye signals into a sequence.
   - nuances of the sequencing chemistry and expert knowledge can be programmed into the sequence analysis software.

7.2 Polymerase Chain Reaction (PCR)

PCR is an *in vitro* technique for the amplification of a region of DNA which lies between two regions of known sequence. PCR amplification is achieved by using oligonucleotide primers. These are typically short, single stranded oligonucleotides which are complementary to the outer regions of known sequence.

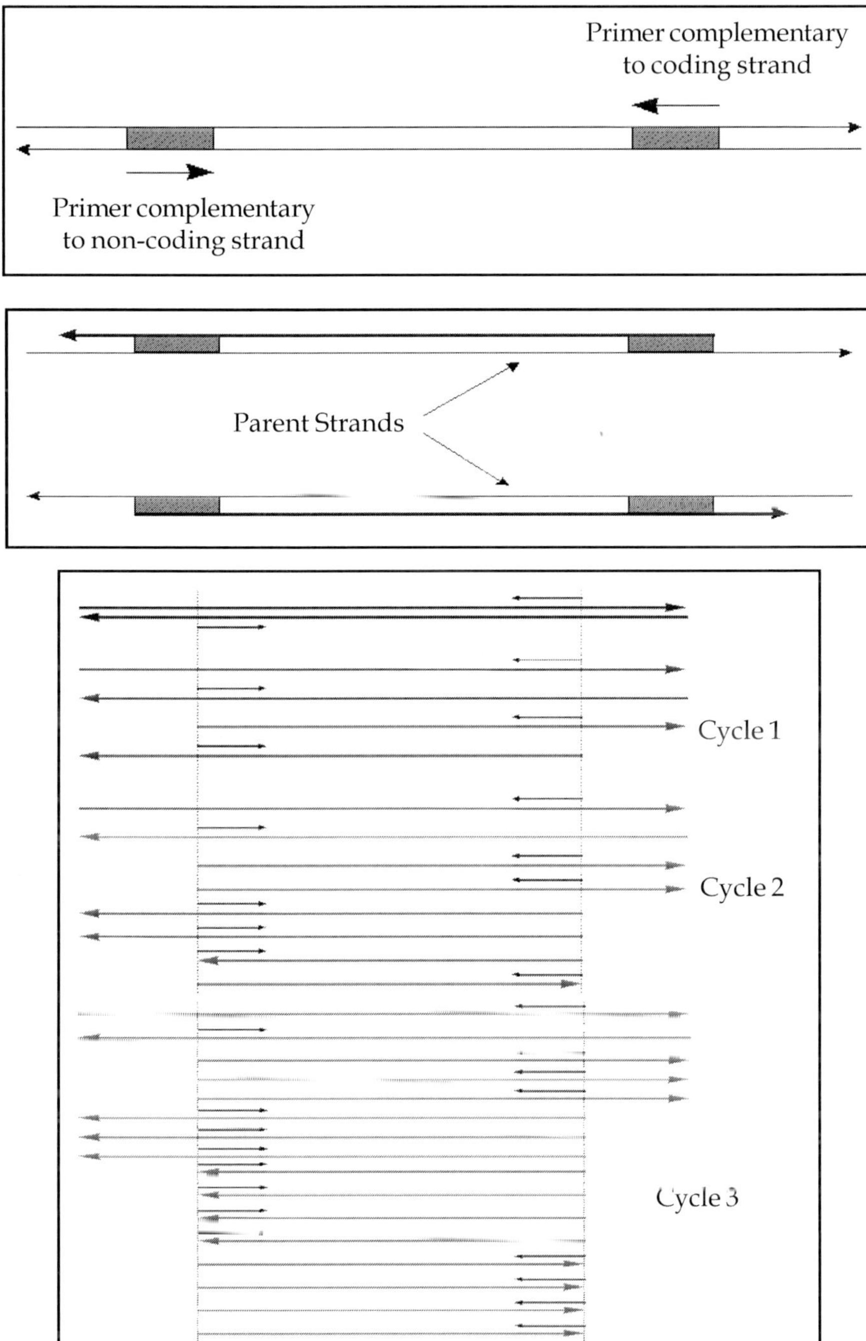

- The oligonucleotides serve as primers for DNA polymerase and the denatured strands of the large DNA fragment serves as the template.
  - This results in the synthesis of new DNA strands which are complementary to the parent template strands.
  - These new strands have defined 5' ends (the 5' ends of the oligonucleotide primers), whereas the 3' ends are potentially ambiguous in length.
- The oligonucleotide directed synthesis of daughter DNA strands can be repeated if the new duplex is denatured (by heating) and additional primers are allowed to anneal (by cooling to an appropriate temperature).

The steps of:

1. Template denaturation
2. Primer annealing
3. Primer extension

Comprise a single "cycle" in the PCR amplification methodology.

After each cycle the newly synthesized DNA strands can serve as templates in the next cycle.

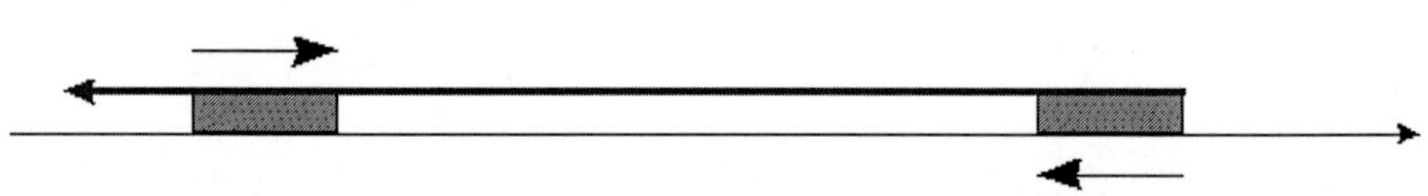

Only showing half the products of the first elongation reaction

Note that half of the newly synthesized strands from the second round of replication have 5' and 3' termini that are defined by the annealing location of the primer oligonucleotides.

Summary of products at the end of each PCR cycle:

The total of each of the different types of oligonucleotide fragments for each cycle can be summarized as :

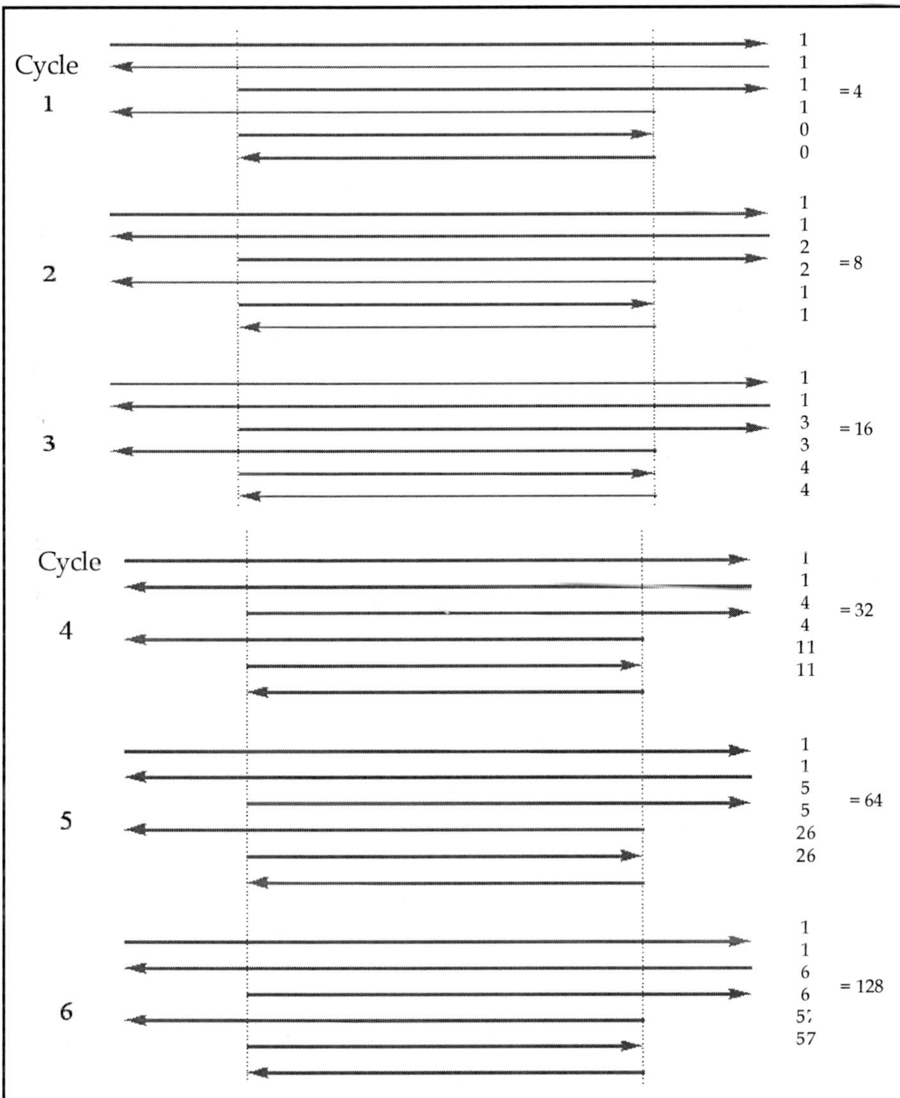

The amplification of the fragments follows the following pattern:

1. There is always 1x copy of each of the original templates - the PCR as outlined never reproduces the full length template

2. There will be 'n' x copies of each fragment with an indeterminate length (where 'n' is the number of cycles). The fragments of indeterminant length have one end defined by a PCR primer and the other end is of indeterminant length

3. There will be $(2^n - (n + 1))$ x copies of each fragment of defined length (i.e. each end defined by the two PCR primers).

- For example, after 6 cycles we will have
  - 1 x copy of each original template
  - 6 x copies of each fragment of indeterminant length
  - (64 - (6 + 1)) = 57 x copies of each fragment of defined length
- In the above case, the desired PCR product will be a duplex of the defined length fragment. The question is: how many will be produced?
  - The original templates will not necessarily anneal with one another
  - Likewise, the templates of indeterminant length will not necessarily anneal with one another
  - Furthermore, as the number of cycles proceeds, the defined length fragments far outnumber the others
  - Therefore, the original template, and fragments of indeterminant length are most likely going to hybridize with defined length fragments
  - Our expected amplification would therefore be (using the 6 cycle PCR example)

57 - (6 + 1) = 50 x copies of each defined length fragments (i.e. 50 x duplexes of defined length)

- The expected *amplification* of the desired defined length product with respect to the original template concentration 'x' can thus be represented by the formula: $[(2^n - (n + 1)) - (n + 1)]$ x or $(2^n - 2(n + 1))$ (this is often abbreviated to a simple rule of thumb for the amplification: $(2^n - 2n)$ x)
- The interpretation of this formula is that
  - For a given number of cycles 'n' we make '2^n x' *total possible duplexes.*
  - For a given number of cycles there will be '2(n+1) (or 2n in our approximation) x' duplexes, which are, formed from either the original template, or a fragment of indeterminate length, along with a fragment of defined length.

  - Thus, the total concentration of desired product (duplexes with a length defined by the PCR primers) will be $(2^n - 2(n+1))$ x (where x is the concentration of the original duplex).

- The theoretical amplification value is never achieved in practice. Several factors prevent this from occuring, including:

1. Competition of complementary daughter strands with primers for reannealing (i.e. two daughter strands reannealing results in no amplification).
2. Loss of enzyme activity due to thermal denaturation, especially in the later cycles.
3. Even without thermal denaturation, the amount of enzyme becomes limiting due to molar target excess in later cycles (i.e. after 25 - 30 cycles too many primers need extending)
4. Possible second site primer annealing and non-productive priming.

PCR was invented in 1985 by Kary Mullis, working for Cetus corporation somewhere near Berkeley, California. The original method of PCR used the Klenow fragment *of E. coli* DNA polymerase I. This enzyme, however, denatures at temperatures lower than that required to denature most template duplexes. Thus, after each cycle, fresh enzyme had to be added to the reaction. This was quite tedious. In addition to this problem with the enzyme, the samples had to be moved from one temperature bath to another to allow the individual steps of denaturation, annealing and polymerization (which all required different temperatures. This was pretty tedious too.

Thus, two main advances allowed the process to be automated, these advances were:

1. The use of thermostable DNA polymerases, which resisted denaturation (inactivation) at high temperatures. Thus, an initial aliquot of polymerase could last throughout the numerous cycles of the protocol. The first thermostable DNA polymerase to be used was isolated from the bacterium *Thermus aquaticus*. It was isolated from a hot spring in Yellowstone National Park where it lived happily (i.e. it replicated its DNA) at temperatures in excess of 85 °C.

2. The development of temperature baths, which could shift their temperatures up and down rapidly and in an automated, programmed manner. These are known as thermal cyclers or PCR machines.

Thermal cycling parameters

The thermal cycling parameters are critical to a successful PCR experiment. The important steps in each cycles of PCR include:

1. Denaturation of template
2. Annealing of primers
3. Extension of the primers

A representative temperature profile for each cycle might look like the following:

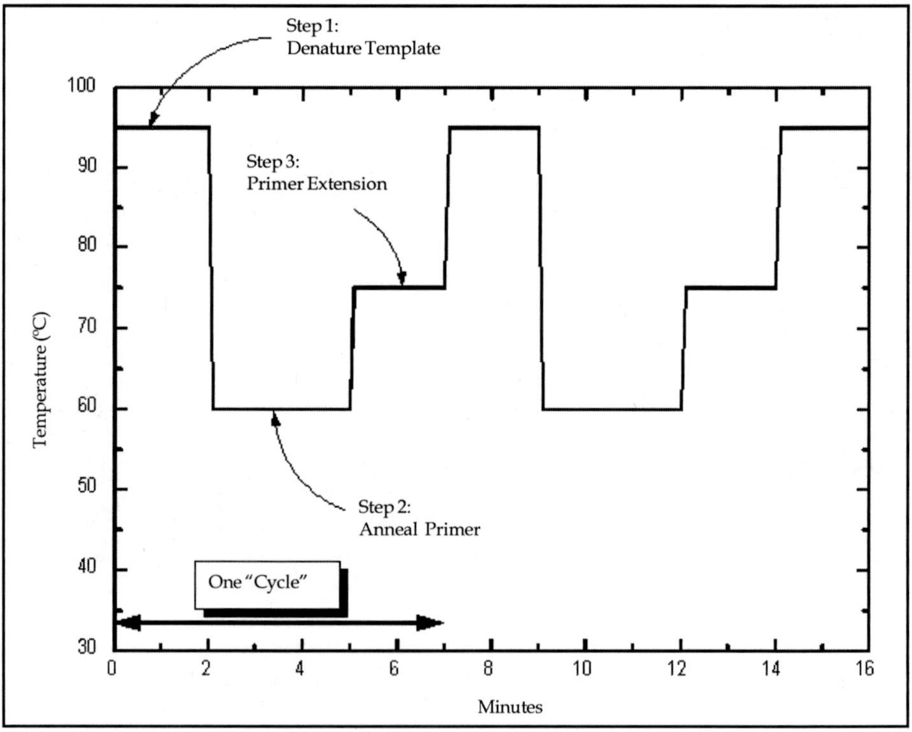

Template denaturation

The initial denaturation of template is accomplished at 95-100 °C.

- Supercoiled plasmids are tougher to melt and may require boiling for several minutes, or may be initially denatured by using base (sodium hydroxide, followed by pH neutralization).
- Denaturation during the PCR experiment (i.e. second cycle onward) is usually accomplished at temperatures of 92-95 °C.

Primer annealing temperature

Primer annealing temperature is an important parameter in the success of the PCR experiment. The annealing temperature is characteristic for each oligonucleotide:

- It is a function of the length and base composition of the primer as well as the ionic strength of the reaction buffer.
- Estimates of the annealing temperature can be calculated in several different ways.
- These calculated annealing temperatures are a *starting point* for the PCR experiment, but ideal annealing temperatures are determined empirically.

Primer extension

Primer extension is usually performed at 72 °C, or the *optimum temperature of the DNA polymerase*. The length of time of the primer extension steps can be increased if the region of DNA to be amplified is long, however, for the majority of PCR experiments an extension time of 2 minutes is sufficient to get complete extension.

Number of cycles

The number of cycles is usually between 25 and 35.

- More cycles mean a greater yield of product.
- However, with increasing number of cycles the greater the probability of generating various artifacts (e.g. mispriming products).
- It is unusual to find procedures which have more than 40 cycles.

Choice of Polymerases for PCR

One of the important advances which allowed development of PCR was the availability of thermostable polymerases. This allowed initially added enzyme to survive temperature cycles approaching 100 °C.

Thermostable DNA polymerases and their sources

DNA Polymerase	Natural or recombinant	Source
Taq	Natural	*Thermus aquaticus*
Amplitaq®	Recombinant	*T. aquaticus*
Amplitaq (Stoffel fragment)®	Recombinant	*T. aquaticus*
Hot *Tub*™	Natural	*Thermus flavis*
Pyrostase™	Natural	*T. flavis*
Vent™	Recombinant	*Thermococcus litoralis*
Deep Vent™	Recombinant	*Pyrococcus GB-D*
Tth	Recombinant	*Thermus thermophilus*
Pfu	Natural	*Pyrococcus furiosus*
ULTma™	Recombinant	*Thermotoga maritima*

Properties of DNA polymerases used in PCR

	Taq/ Amplitaq®	**Stoffel fragment**	**Vent™**	**Deep Vent™**	**Pfu**	**Tth**	**ULTma™**
95 °C half-life	40 min	80 min	400 min	1380 min	>120 min	20 min	>50 min
5′3′ exo	+					+	
3′5′ exo			+	+	+		+
Extension rate (nt/sec)	75	>50	>80	?	60	>33	?
RT activity	Weak	Weak	?	?	?	Yes	?
Resulting ends	3' A	3' A	>95% blunt	>95% blunt	blunt	3' A	blunt
Strand displacement			+	+			
M.W. (kDa)	94	61	?	?	92	94	70

Buffers and $MgCl_2$ in PCR reactions

A typical reaction buffer for PCR would something like:

- 10 mM Tris, pH 8.3
- 50 mM KCl
- 1.5 mM $MgCl_2$
- 0.01% gelatin
- The $MgCl_2$ concentration in the final reaction mixture is usually between 0.5 to 5.0 mM, and the optimum concentration is determined empirically (typically between 1.0 - 1.5 mM). Mg^{2+} ions:

Form a soluble complex with dNTP's which is essential for dNTP incorporation, stimulate polymerase activity and increase the Tm (melting temperature) of primer/template interaction (i.e. it serves to stabilize the duplex interaction.

Generally,

- low Mg^{2+} leads to low yields
- high Mg^{2+} leads to accumulation of nonspecific products (mispriming).

Primers

Primer design

- Primers used are 20 - 30 mer in length. This provides for practical annealing temperatures.
- Primers should avoid stretches of polybase sequences (e.g. poly dG) or repeating motifs - these can hybridize with inappropriate register on the template.
- Inverted repeat sequences should be avoided so as to prevent formation of secondary structure in the primer, which would prevent hybridization to template
- Sequences complementary to other primers used in the PCR should be avoid so as to prevent hybridization between primers (particularly important for the 3' end of the primer)
- If possible the 3' end of the primer should be rich in G, C bases to enhance annealing of the end which will be extended

- The distance between primers should be less than 10 Kb in length. Typically, substantial reduction in yield is observed when the primers extend from each other beyond ~3 Kb.

Melting temperature (Tm) of primers

The melting temperature of primer hybridization can be calculated using various formulas. The most commonly used formula is as under:

(1) Tm = [(number of A+T residues) x 2 °C] + [(number of G+C residues) x 4 °C]

- This formula was determined originally from oligonucleotide hybridization assays, which were performed in 1 M NaCl, and appears to be accurate in lower salt conditions only for primers less than about 20 nucleotides in length.
- The common wisdom is that the Tm is more like 3-5 °C lower than the value calculated from this formula.

(2) Tms = 81.5 + 16.6(log10[J+]) + 0.41(%G+C) - (600/l)

- Where $[J^+]$ = the molar concentration of monovalent cations (e.g. Na^+ from NaCl), and l = the length of oligonucleotide. (%G+C) is the actual percentage value and not a fractional representation (i.e. the value to insert for a primer which had 90 % G+C content would be "90" and not "0.90").
- This formula is reportedly useful for primers of 14 to 70 bases in length.

(3) Tmp = 22 + 1.46([2 x (G+C)] + (A+T))

- This formula is reportedly useful for primers of 20-35 bases in length.
- The calculated annealing temperature is only a reference temperature from which to initiate experiments.
- The actual annealing temperature may be 3-12 °C higher than the calculated Tm.
- The actual annealing temperature condition should be determined empirically.
- The highest annealing temperature which gives the best PCR product should be used.
- Examples of T_m, T_ms and T_mp calculations (0.05 M K^+)

	G	A	T	C	T_m	T_ms	T_mp
15 mer	3	5	2	5	46	42	56
20 mer	6	5	4	5	62	52	67
30 mer	8	6	8	8	92	62	89

Calculating primer concentrations

The molar concentration of a primer can be calculated based on the absorbance of the primer at 260 nm (A_{260}) and the molar extinction coefficient for the primer at this wavelength. The molar extinction coefficient for the primer can be calculated by knowing the sequence of the primer and then summing the molar extinction values for the individual bases which comprise the primer. The individual bases have the following molar extinction coefficients at 260 nm:

- A 1.0 molar solution of dT has a value of 8,400 absorbance units at 260 nm.
- A 1.0 molar solution of dA has a value of 15,200 absorbance units at 260 nm.
- A 1.0 molar solution of dG has a value of 12,010 absorbance units at 260 nm.
- A 1.0 molar solution of dC has a value of 7050 absorbance units at 260 nm.
- For example, the primer 5' TAGC 3' would have a molar extinction coefficient of 42,660 at 260 nm. Likewise, a 10 micromolar solution of this primer would give an absorbence of 0.427 at 260 nm.

7.3 Restriction Fragment Length Polymorphisms (RFLPs)

The first types of DNA markers that were used for genetic mapping were RFLPs. The DNA sequence differences between polymorphic lines may create differences in the length of restriction fragments derived from genomic DNA. For instance a given restriction site may be present in one line and not in the other. As illustrated in Figure 1, this polymorphism can be revealed by genomic DNA blot hybridisation (Southern) using as probe a DNA fragment corresponding to that region. The polymorphic bands can then be used as genetic markers to distinguish the two lines. Multiple RFLP markers can be identified and assembled into genetic maps. An advantage of the RFLP mapping approach is that RFLP markers are co-dominant. Distinct patterns are indeed obtained for plants that are homozygous or heterozygous for the parental alleles (Figure 1). Hence, all the chromosomes of a given F2 population can be scored. In contrast, a main disadvantage is that RFLP mapping necessitates relatively large amounts of DNA because distinct RFLP markers may require digestion of genomic DNA samples with different diagnostic restriction enzymes.

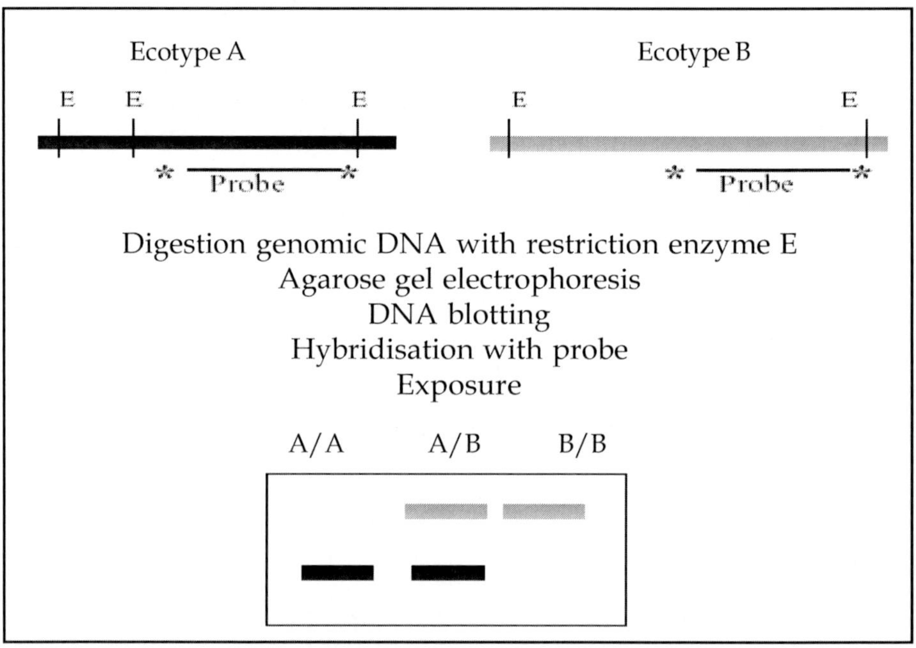

Figure 1. Principle of RFLP markers.

This figure illustrates an RFLP marker which utilises a site for the restriction enzyme (E) which is present in line A and not in line B.

7.4 Random Amplification of Polymorphic DNA (RAPD)

RAPD reactions are PCR reactions, but they amplify segments of DNA which are essentially unknown to the scientist (random).

PCR is used to amplify a known sequence of DNA. Thus, the scientists chooses the sequence he or she wants to amplify, then designs and makes primers which will anneal to sequences flanking the sequence of interest. Thus, PCR leads to the amplification of a particular segment of DNA.

Standard PCR

However, in RAPD analysis, the target sequence is unknown. The scientist will design a primer with an arbitrary sequence. In other words, the scientist simply makes up a 10 base pair sequence, then synthesizes the primer. The scientist then carries out a PCR reaction and runs an agarose gel to see if any DNA segments were amplified in the presence of the arbitrary primer.

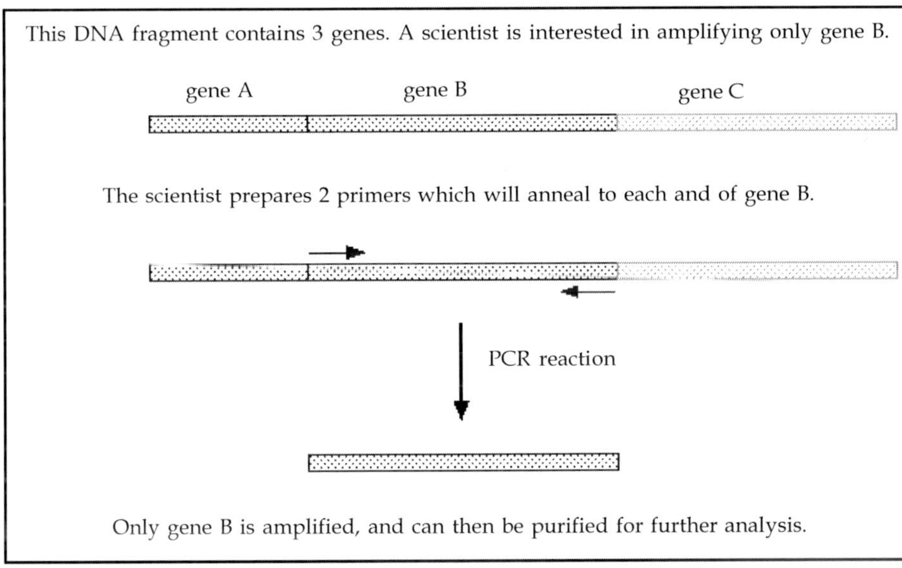

In order for PCR to occur

1. The primers must anneal in a particular orientation .
2. The primers must anneal within a reasonable distance of one another.

In this figure which depicts a RAPD reaction, a large fragment of DNA is used the template in a PCR reaction containing many copies of a single arbitrary primer.

RAPD Reaction 1

1. The arrows represent multiple copies of a primer (all primers (arrows) have the same sequence). The direction of the arrow also indicates the direction in which DNA synthesis will occur.
2. The numbers represent locations on the DNA template to which the primers anneal.

3. Primers anneal to sites 1, 2, and 3 on the bottom strand of the DNA template and primers anneal to sites 4, 5, and 6 on the top strand of the DNA template.

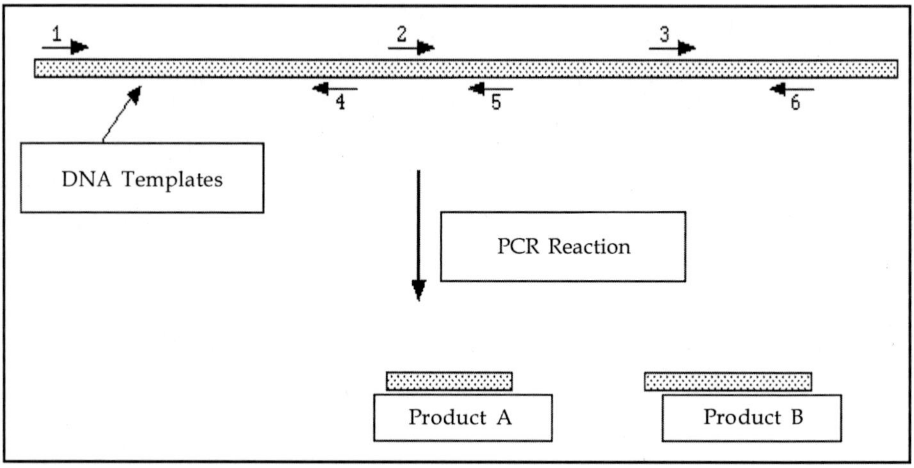

2 RAPD PCR products are formed (example)

1) Product A is produced by PCR amplification of the DNA sequence which lies in between the primers bound at positions 2 and 5.

2) Product B is the produced by PCR amplification of the DNA sequence which lies in between the primers bound at positions 3 and 6.

Note that no PCR product is produced by the primers bound at positions 1 and 4 because these primers are too far apart to allow completion of the PCR reaction. Also note that no PCR products are produced by the primers bound at positions 4 and 2 or positions 5 and 3 because these primer pairs are not oriented towards each other.

Finding Differences Between Genomes Using RAPD Analysis

If another DNA template (genome) was obtained from a different (yet related) source, there would probably be some differences in the DNA sequence of the two templates.

Suppose there was a change in sequence at primer annealing site 2:

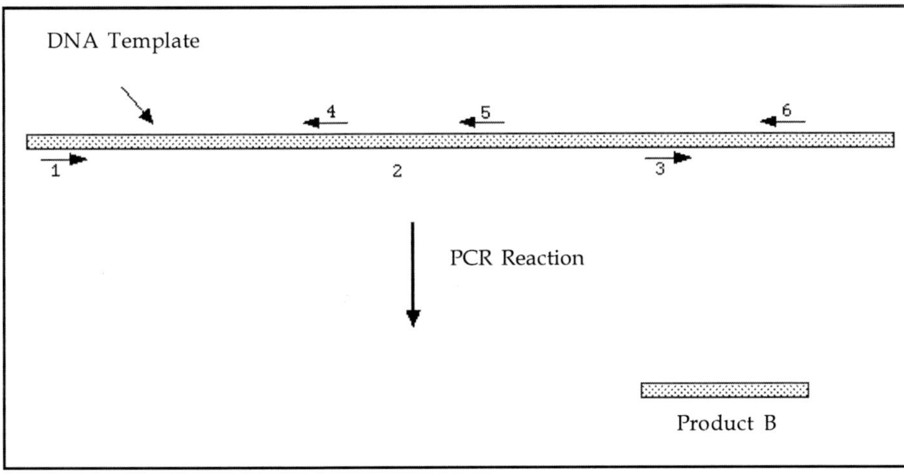

RAPD Reaction 2

As shown in this figure, the primer is no longer able to anneal to site 2, and thus the PCR product A is not produced. Only product B is produced. If you were to run the 2 RAPD PCR reactions diagramed above on an agarose gel

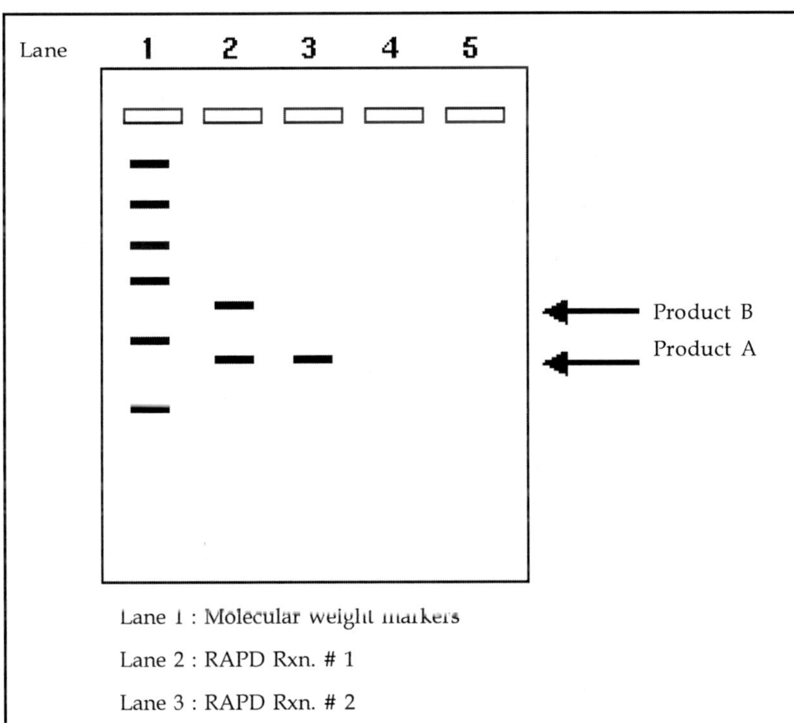

7.5 Amplified Restriction Fragment Length Polymorphism (AFLP)

Amplified Fragment Length Polymorphism. A hybrid of RFLP and RAPD techniques. Genomic DNA is cut with restriction enzymes, as in RFLP. Typically, two different restriction enzymes are used. The idea is to produce a large number of fragments. Some of the fragments are selectively amplified with PCR using "random" primers, as in RAPD. The primers are not really random, however. Specific oligonucleotide "adapters" (these are complementary to the restriction sites) of 25-30 bp are ligated to the restricted DNA fragments. The primers are complementary to these adapters. However, the primers vary at their 3'-end, such that they will amplify only a subset of the restricted DNA fragments. Typically, a moderate-to-high number of bands will be observed, most of which may be monomorphic. Polymorphic bands are identified as in RAPD analysis. Polymorphic bands can even be excised from the gel and sequenced. This allows for specific PCR primers to be developed, if necessary. AFLPs have typically been used to study variation among individuals of a species, most commonly for producing genetic maps (and in trying to find genes responsible for certain traits). They have received limited attention as tools in systematics, perhaps because the method is relatively labor intensive when compared with other methods. The power of AFLP is based upon the molecular genetic variations that exist between closely related species, varieties, or cultivars. These variations in DNA sequence are exploited by the AFLP technology such that "fingerprints" of particular genotypes can be routinely generated. These "fingerprints" are simply RFLPs visualized by selective PCR amplification of DNA restriction fragments.

❑❑❑

8

DNA Cloning

DNA cloning facilitates the isolation and manipulation of fragments of an organism's genome by replicating them independently as a part of an autonomous vector. In general detailed molecular analysis of protein or other constituents of most organisms was very difficult or impossible by their scarcity and the consequence difficulty of their purification in large quantities. The substitution approach is to be isolating the genes responsible for specific characters or a formation of specific biological products. Although the organisms genome is very large and complex and any sequence of our interest usually occurs only ones or twice per cell. Hence standard chemical or biochemical methods can not be used to isolate a specific region of genomic DNA for study. The solution of this problem is to place a relatively short fragment of genomic DNA which might contain the gene of interest, in an autonomously replicating piece of DNA known as vector, forming recombinant DNA which can be replicated independently of the original genomes and normally in another host species altogether. Propagation of the host organism containing the recombinant DNA forms a genetically identical organisms known as clone. This process is known as DNA cloning. A number of DNA cloning application are collected together under the term genetic engineering.

Host and vector

Most of the routine manipulation involves in gene cloning use *Escherichia coli* as the host organism. Plasmid and bacteriophase may be used as cloning vector in Escherichia coli. Vector based on plasmids, virus and whole chromosomes have been used to carry foreign genes in to other prokaryotic and eukaryotic organisms. Vector must normally be capable of being replicated and isolated independently of the host genome, although some are designed to incorporate DNA in to the host genome for longer-term expression and cloning genes. It also incorporates a selectable marker, usually conferring resistance to toxin.

8.1 Sub cloning

The simplest- kind of cloning experiment, which exemplified many of the basic techniques of DNA cloning, is the transfer of a fragment of cloned DNA from one vector to another, a processs known as sub-cloning. This might be used to investigate a short region of a large cloned fragment in more detail, or to transfer a gene to vector design to express it in a particular species.

Steps of sub-cloning

1. Isolation of DNA containing sequence of interest.
2. Digestion with restriction endonuclease to form discrete fragment
3. Seporation of fragment with gel electrophoresis.
4. Purification of target fragment.
5. Ligation of fragment with vector.
6. Transfer of vector in host cell.
7. Selection of transformed bacteria/host cell.
8. Analysis of recombinant plasmid.

Cloning into a Plasmid

Many diseases are caused by gene alterations. Our understanding of genetic diseases was greatly increased by information gained from DNA cloning. In DNA cloning procedure, a DNA fragment that contains a gene of interest is inserted into a cloning vector or plasmid.

The plasmid carrying genes for antibiotic resistance, and a DNA strand, which contains the gene of interest, are both cut with the same restriction endonuclease. The plasmid is opened up and the gene is freed from its parent DNA strand. They have complementary "sticky ends." The opened plasmid and the freed gene are mixed with DNA ligase, which reforms the two pieces as recombinant DNA.

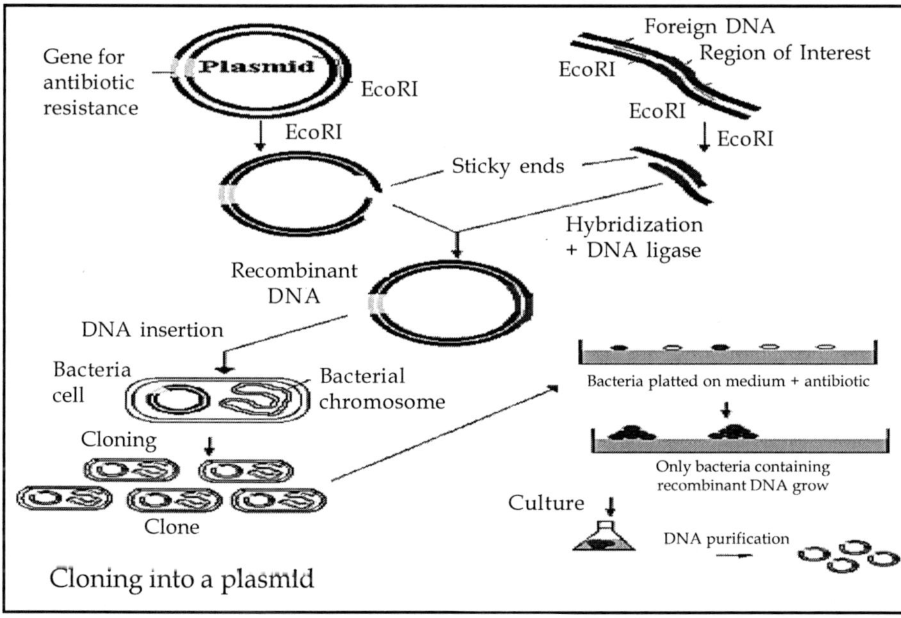

Fig. A plasmid is used to import recombinant DNA into a host cell for cloning.

Plasmids and copies of the DNA fragment produce quantities of recombinant DNA.

This recombinant DNA stew is allowed to transform a bacterial culture, which is then exposed to antibiotics. All the cells except those which have been encoded by the plasmid DNA recombinant are killed, leaving a cell culture containing the desired recombinant DNA.

DNA cloning allows a copy of any specific part of a DNA (or RNA) sequence to be selected among many others and produced in an unlimited amount. This technique is the first stage of most of the genetic engineering experiments: production of DNA libraries, PCR, DNA sequencing.

8.2 Cloning into a Yeast Artificial Chromosome (YAC)

DNA cloning is a technique that allows the wholesale production of a specific DNA sequence. DNA containing a gene of interest is inserted into the purified DNA genome of a self replicating element, which can be a plasmid, a virus or in this case, a yeast artificial chromosome (YAC). A YAC can be considered as a functional artificial chromosome (self replicating element), since it includes three specific DNA sequences that enable it to propagate from one cell to its offspring:

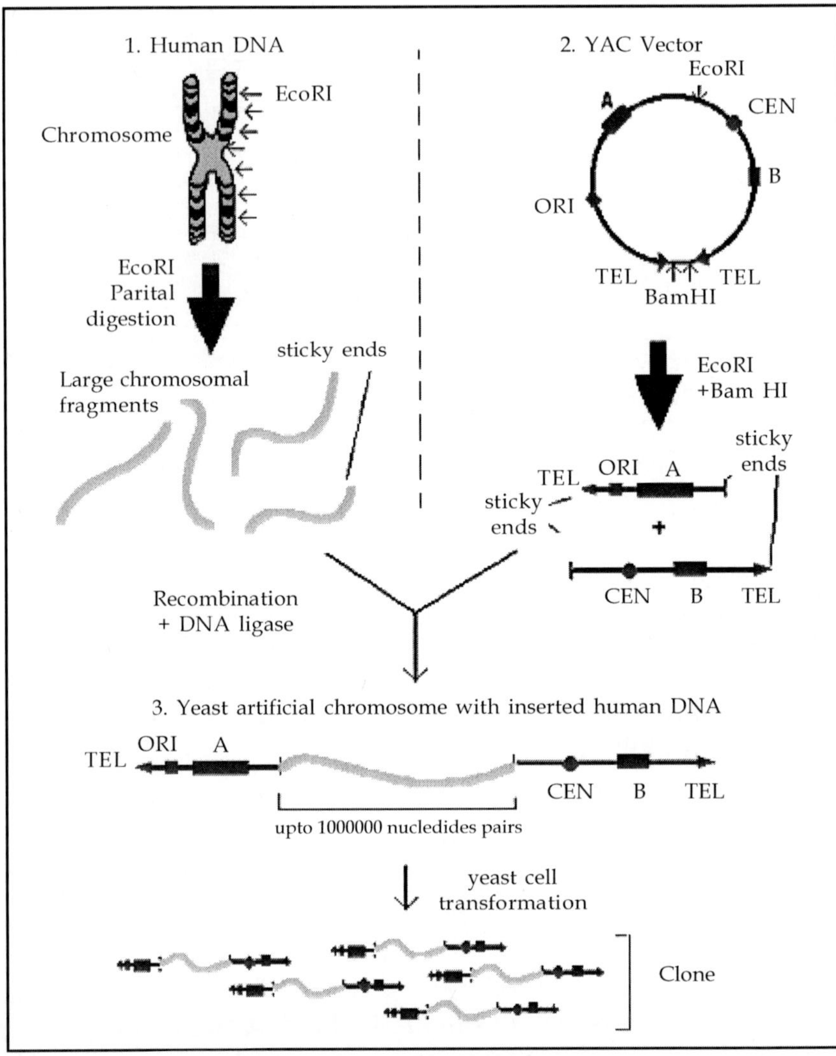

Fig. Cloning into a Yeast Artificial Chromosome (YAC)

1. The telomere, which is located at each chromosome end, protects the linear DNA from degradation by nucleases.
2. The centromere that is the attachment site for mitotic spindle fibers, "pulls" one copy of each duplicated chromosome into each new daughter cell.
3. Replication origin sequences, which are specific DNA sequences, those allow the DNA replication machinery to assemble on the DNA and move at the replication forks.

4. A and B: selectable markers that allow the easy isolation of yeast cells that have taken up the artificial chromosome.
5. Recognition site for the two restriction enzymes EcoRI and BamHI.

DNA cloning into a plasmid allows the insertion of DNA fragment of about 10,000 nucleotide base pairs, DNA cloning into a YAC allows the insertion of DNA fragments up to 1,000,000 nucleotide base pairs. it is important to map the entire human genome (3x1,000,000,000 nucleotide base pairs) it would require more than 100,000 plasmid clones. In principle, the human genome could be represented in about 10,000 YAC clones.

Techniques for cloning genomic DNA into yeast artificial chromosomes (YAC) make it possible to analyze very long DNA sequences like human genes.

Cloning human genomic DNA into a YAC

1. Genomic DNA is partially digested by the restriction enzyme EcoRI. Very large DNA fragments are obtained.
2. The YAC is digested by the two restriction enzymes EcoRI and BamHI.
3. Those two elements recombine at the EcoRI sites and are covalently linked by the DNA ligase.
4. A recombinant YAC vector, a yeast artificial chromosome with genomic DNA inserted, is produced. This vector can be used to infect yeast cells and generated an unlimited number of copies.

8.3 DNA libraries

There are two main source from which DNA is derived for cloning experiment design ton identify an un known gene- bulk genomic DNA from yhe species of interest. And, in the case of eukaryotes, bulk mRNA from a cell or tissue where the gene is known to express. They are used in the formation of genomic library and cDNA libraries. They are set of DNA clones (a clones is a genetically distinct individual or set of identical individuals.), each of which has been derived from the insertion of a different fragment into a vector followed by propagation in the host. Genomic libraries are prepared from random fragments of genomic DNA. How ever the genomic library are in efficient method of finding a gene, particularly in large eukaryotic genomes, where much of DNA is non-coding. The alternative is to use as the source of the library the mRNA from the cell or tissue. Which is known to express the gene. DNA copies (cDNA) are synthesized from the mRNA by reverse transcription and are then inserted into a vector to form a cDNA library. cDNA libraries are efficient for cloning a gene sequence, but yield only the coding region, and not the surrounding non-coding genomic DNA sequence.

Screening libraries

Since it is not apparent which clone in a library contains the gene of interest, a method for screening for its presence is required. This is often based on the use of a radioactively labled probe which is complementary or partially complementary to a region of the gene sequence, and which can be used to detact it by hybridization. The probe sequence might be an oligonucleotide derived from the sequence of the product of a gene, if it is available, or form a related gene from another species. An increasingly important method for the generation of probe is the polymarase chain reaction (PCR). Other screening methods rely on the expression of the coding sequence of the clones in the library, and identification of the protein product from its activity.

Analysis of clone

Once a clone containing a target DNA / gene is identified, the structure of the cloned fragment may be investigated further using restriction mapping, the analysis of a fragmentaton of the DNA with restriction endonuclease or ultimately by the sequencing of entire fragment of the DNA. The sequence can than be analyzed by comparison with other known sequence from databases and the complete sequence of the gene determine. The sequence is then too available for manipulation in any of the applications of cloning.

❑❑❑

9

ENZYMES DEALS WITH DNA

9.1 Methylases

- The recognition sequences of the methylases are the same as the associated endonucleases (e.g. EcoR1 methylase recognizes and methylates at the sequence "GAATTC").
- All methylases transfer the methyl group from S-adenosyl-methionine (SAM) to a specific base in the recognition sequence, and SAM is a required component in the methylation reaction.
- Methylation of DNA usually has the effect of protecting the DNA from the related restriction endonuclease. However, there are methylases with minimal specificity. For example, Sss I methylase will methylate cytosine residues in the sequence 5' ... CG ... 3'. In this case, the methylated DNA will be protected from a wide variety of restriction endonucleases.
- Some restriction endonucleases will only cut DNA at their recognition sites if the DNA is methylated (e.g. Dpn I).

Still other restriction endonucleases will cut both methylated and non-methylated DNA at their recognition sequences (e.g. BamH I).

9.2 Polymerases

- They add nucleotides to the 3'-OH end of a primer
- The order of the nucleotides in the nascent polynucleotide is template directed
- In addition to the 5'→3' polymerase activity, polymerases can contain exonuclease activity. This exonuclease activity can proceed *either* in the 5'→3' direction, or in the 3'→5' direction.
- Exonuclease activity in the 3'→5' direction allows the polymerase to correct a mistake if it incorporates an incorrect nucleotide (so called "error correction activity"). It can also slowly degrade the 3' end of the primer.
- Exonuclease activity in the 5'→3' direction will allow it to degrade any other hybridized primer it may encounter. Without 5'→3' exonuclease activity, obstructing primers may or may not be physically deplaced, depending on the polymerase being used.

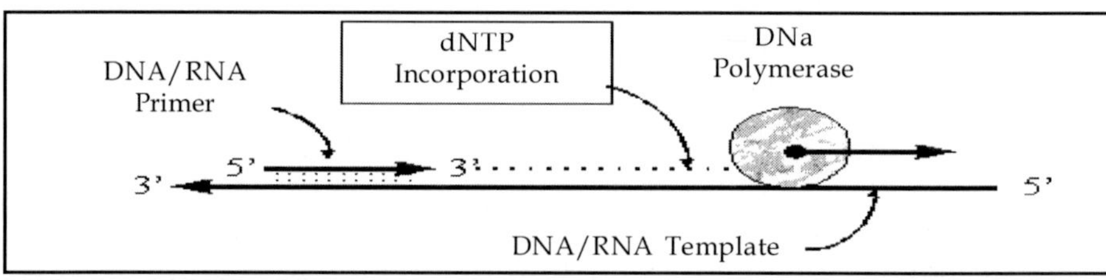

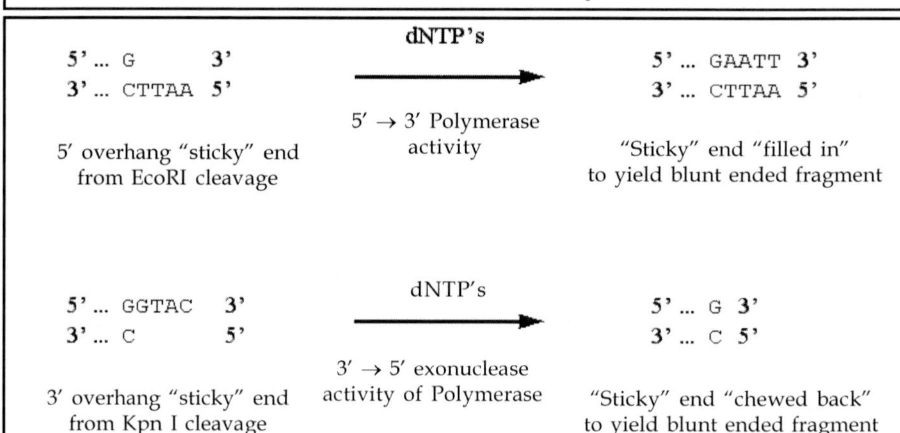

Use of Polymareses

The various activities of the different polymerases lend them to a variety of applications. For example, restriction endonucleases can yield fragments of DNA with either 3' or 5' nucleotide "overhangs".

- In the case of 5' overhangs, the 5'→3' polymerase activity can fill these in to make blunt ends.
- In the case of 3' overhangs, the 3'→5' exonuclease activity present in some polymerases (especially T4 DNA polymerase) can "chew back" these ends to also make blunt-ended DNA fragments.

"Nick-translation"

This method is used to obtain highly radiolabeled single strand DNA fragments, which makes use of 5'->3' exonuclease activity present in some polymerases (*E. coli* DNA polymerase I, for example). In this method a DNA duplex of interest is "nicked" (i.e. one of the strands is cut; see DNAse I).

Then DNA pol I is added along with radiolabeled nucleotides. The 5'->3' exonuclease activity chews away the 5' end at the "nick" site and the polymerase activity incorporates the radiolabeled nucleotides. The resulting polynucleotide will be highly radiolabeled and will hybridize to the DNA.

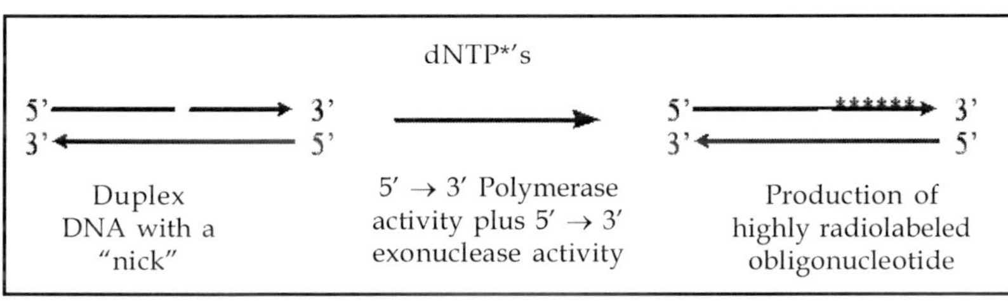

Thermostable polymerases have the ability to remain functional at temperature ranges where the DNA duplex will actually "melt" and become separated. This has allowed the development of the "Polymerase Chain Reaction" technique (PCR), which has had a profound impact on modern Biotechnology.

9.3 Nucleases

Exonuclease

Exonuclease will degrade both 3' and 5' termini of double stranded DNA. It will not make internal cleavages ("nicks"), however, it will degrade the ends of DNA at existing internal "nicks" . The degradation of termini is not coordinated, meaning that the product is not 100% blunt ended.

Such ends can be made blunt by filling in and chewing back by a suitable polymerase (e.g. T4 DNA polymerase). The unit definition is 1 unit is amount of enzyme required to remove 200 base pairs from each end of duplex DNA in 10 minutes at 30 °C.

Exonuclease III

- Catalyzes the stepwise removal of nucleotides from the 3' hydroxyl termini of duplex DNA.
- The enzyme will attack the 3' hydroxyl at duplex DNA with blunt ends, with 5' overhangs, or with internal "nicks".
- Since duplex DNA is required, the enzyme will not digest the 3' end of duplex DNA where the termini are 3' overhangs.

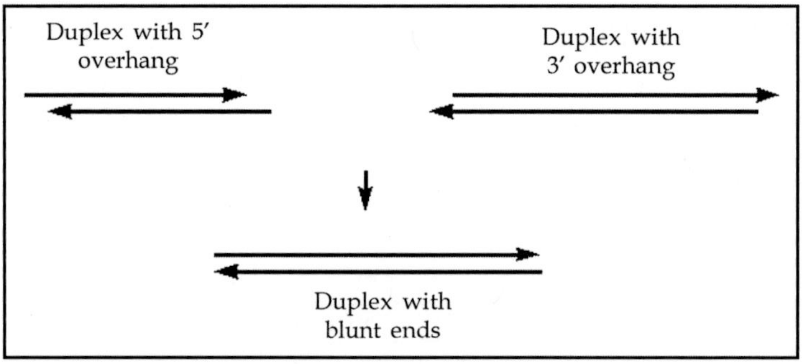

Deoxyribonuclease I (DNAse I)

This enzyme hydrolyzes duplex or single DNA strands preferentially at the phosphodiester bonds 5' to pyrimidine nucleotides. In the presence of Mg^{2+} ion, DNAse I attacks each strand independently and produces nicks in a random fashion and it could be useful for nick-translation. In the presence of Mn^{2+} ion DNAse I cleaves both strands of DNA at approximately the same position.

9.4 Ligases

Ligases catalyze the formation of a phosphodiester bond between juxtaposed 5' phosphate and 3' hydroxyl termini of nucleotides (potentially RNA or DNA depending on the ligase). In a sense, they are the opposite of restriction endonucleases, but they do not appear to be influenced by the local sequence, *per se*. Ligases require either rATP or NAD+ as a cofactor, and this contrasts with restriction endonucleases.

The following are different types of ligases and their characteristics.

T4 DNA ligase

1. Isolated from bacteriophage T4.
2. Will ligate the ends of duplex DNA or RNA.
3. This enzyme will join blunt-end termini as well as ends with cohesive (complementary) overhanging ends (either 3' or 5' complementary overhangs).
4. This enzyme will also repair single stranded nicks in duplex DNA, RNA or DNA/RNA duplexes. Requires ATP as a cofactor.

Taq DNA ligase

- This ligase will catalyze a phosphodiester bond between two adjacent oligonucleotides which are hybridized to a complementary DNA strand:

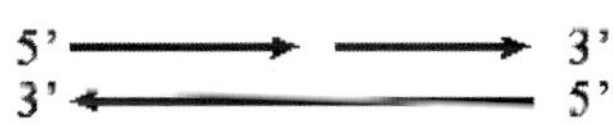

- The ligation is efficient only if the oligonucleotides hybridize perfectly with the template strand.
- The enzyme is active at relatively high temperatures (45 - 65 °C). Requires NAD+ as a cofactor.

T4 RNA ligase

- It will catalyze formation of a phosphodiester bond between RNA/RNA oligonucleotides, RNA/DNA oligonucleotides, or DNA/DNA oligonucleotides.
- Requires ATP as a cofactor.
- This enzyme does not require a template strand.

T4 RNA ligase can be used for a variety of purposes including constructing RNA/DNA hybrid molecules.

Polynucleotide kinase

- Catalyzes the transfer and exchange of a phosphate group from the α position of rATP (adenine ribose triphosphate nucleotide) to the 5' hydroxyl terminus of double stranded and single stranded DNA or RNA, and nucleoside 3' monophosphates.
- The enzyme will also remove 3' phosphoryl groups.

- Oligonucleotides which are obtained from automated synthesizers lack a 5' phosphate group, and thus, cannot be ligated to other polynucleotides.

Polynucleotide kinase can be used to phosphorylate the 5' end of such polynucleotides.

9.5 Phosphatase

- Catalyzes the removal of 5' phosphate groups from RNA, DNA and ribo- and deoxyribo- nucleoside triphosphates (e.g. ATP, rATP).
- CIP treated duplex DNA cannot self ligate.
- Hemi-phosphorylated duplexes will be ligated on one strand (the phosphorylated strand) and remain "nicked" on the other.

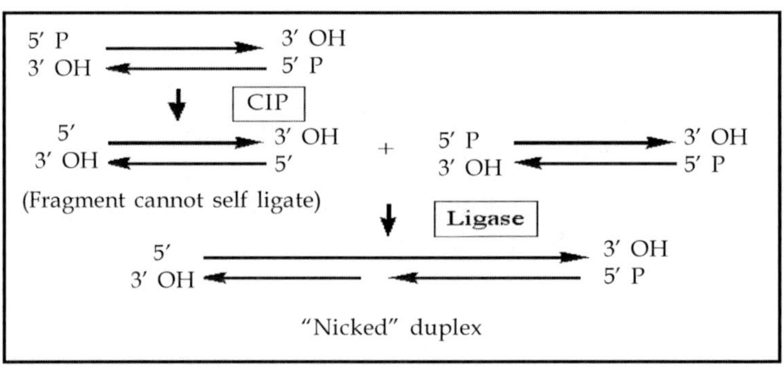

9.6 Topoisomerase

- Topoisomerases cut one strand of the DNA (i.e. it "nicks" the DNA duplex).
- The 5' phosphate of the nicked strand is covalently attached to a tyrosine in the protein.
- The 3' end of the nick then passes once through the duplex.
- *The nick is then resealed, and the linkage number is change by a value of +1.*
- This can therefore result in the removal of a single negative supercoil.

9.7 Restriction enzymes

Restriction enzymes were discovered over 30 years ago during investigations into the phenomenon of host-specific restriction and modification of bacterial viruses. Bacteria initially resist infections by new viruses, and this "restriction" of viral growth stemmed from endonucleases within the cells, which destroy foreign DNA molecules. Among these "restriction enzymes" to be purified were EcoRI and EcoRII from *Escherichia coli*, and HindII and HindIII from *Haemophilus influenzae*. These enzymes were found to cleave DNA at specific sites, generating discrete, gene-size fragments and that may be again joined together in the laboratory. Researchers were quick to recognize that restriction enzymes provided them with a remarkable new tool for investigating gene organization, function as well as expression.

Restriction enzymes were found only within prokaryotes with some exceptions of some viruses. Many thousands of bacteria and archae have now been screened for their presence. Analysis of sequenced prokaryotic genomes indicates that they are common—all free-living bacteria and archaea appear to code for them.

Restriction enzymes are exceedingly varied; they range in size from the diminutive PvuII (157 amino acids) to the giant CjeI (1250 amino acids) and beyond. Among over 3,000 activities that have been purified and characterized, more than 250 different sequence-specificities have been discovered. The search for new specificities continues, both biochemically, by the analysis of cell-extracts, and computationally, by the analysis of sequenced genomes. Although most activities encountered today turn out to be duplicates - isoschizomers - of existing specificities, restriction enzymes with new specificities are found with regularity.

Cloning improves enzyme purity by separating enzymes from contaminating activities present in the same cells. It also improves enzyme yields and greatly simplifies purification, and it provides the genes for sequencing and analysis, and the proteins for x-ray crystallography.

Restriction enzymes protect bacteria from infections by viruses, and it is generally accepted that this is their role in nature. They function as microbial immune systems. When a strain of *E. coli* lacking a restriction enzyme is infected with a virus, most virus particles can initiate a successful infection. The presence of additional enzymes has a multiplicative effect; a cell with four or five independent restriction enzymes could be virtually impregnable.

Restriction enzymes usually occur in combination with one or two modification enzymes (such as DNA-methyl transferase) that protect the cell's own DNA from cleavage by the restriction enzyme. Modification enzymes recognize the same DNA sequence as the restriction enzyme that they accompany, but instead of cleaving the sequence, they methylate one of the bases in each of the DNA strands. The methyl groups protrude into the major groove of DNA at the binding site and prevent the restriction enzyme from acting upon it. Together, a restriction enzyme and its "cognate" modification enzyme(s) form a restriction-modification system. In some restriction-modification systems the restriction enzyme and the modification enzyme(s) are separate proteins that act independently of each other. In other systems, the two activities occur as separate subunits, or as separate domains, of a larger, combined, restriction-and-modification enzyme.

Restriction enzymes are traditionally classified into three types on the basis of subunit composition, cleavage position, sequence-specificity and cofactor-requirements. However, amino acid sequencing has uncovered extraordinary variety among restriction enzymes and revealed that at the molecular level there are many more than three different kinds.

1. Type I

These enzymes are complex, multisubunit, combination restriction-and-modification enzymes that cut DNA at random far from their recognition sequences. Originally thought to be rare, we now know from the analysis of sequenced genomes that they are common. Type I enzymes are of considerable biochemical interest but they have little practical value since they do not produce discrete restriction fragments or distinct gel-banding patterns.

2. Type II

These enzymes cut DNA at defined positions close to or within their recognition sequences. They produce discrete restriction fragments and distinct gel banding patterns, and they are the only class used in the laboratory for DNA analysis and gene cloning. Rather then forming a single family of related proteins, type II enzymes are a collection of unrelated proteins of many different sorts. Type II enzymes frequently differ so utterly in amino acid sequence from one another, and indeed from every other known protein, that they likely arose independently in the course of evolution rather than diverging from common ancestors.

The most common type II enzymes are those like HhaI, HindIII and NotI that cleave DNA within their recognition sequences. Enzymes of this kind are the principal ones available commercially. Most recognize DNA sequences that are symmetric because they bind to DNA as homodimers, but a few, recognize asymmetric DNA sequences because they bind as heterodimers. Some enzymes recognize continuous sequences (e.g.,EcoRI: GAATTC) in which the two half-sites of the recognition sequence are adjacent, while others recognize discontinuous sequences (e.g., BalI: GCCNNNNNGGC) in which the half-sites are separated. Cleavage leaves a 3′-hydroxyl on one side of each cut and a 5′-phosphate on the other. They require only magnesium for activity and the corresponding modification enzymes require only S-adenosylmethionine. They tend to be small, with subunits in the 200-350 amino acid range.

The next most common type II enzymes, also known as 'type IIs" are those like FokI and AIwI that cleave outside of their recognition sequence to one side. These enzymes are intermediate in size, 400-650 amino acids in length, and they recognize sequences that are continuous and asymmetric. They comprise two distinct domains, one for DNA binding, the other for DNA cleavage. They are thought to bind to DNA as monomers for the most part, but to cleave DNA cooperatively, through dimerization of the cleavage domains of adjacent enzyme molecules. For this reason, some type IIs enzymes are much more active on DNA molecules that contain multiple recognition sites.

The third major kind of type II enzyme, more properly referred to as "type IV" are large, combination restriction-and-modification enzymes, 850–1250 amino acids in length, in which the two enzymatic activities reside in the same protein chain. These enzymes cleave outside of their recognition sequences; those that recognize continuous sequences (e.g., AcuI: CTGAAG) cleave on just one side; those that recognize discontinuous sequences (e.g., BcaI: CGANNNNNNTGC) cleave on both sides releasing a small fragment containing the recognition sequence. The amino acid sequences of these enzymes are varied but their organization are consistent. They comprise an N-terminal DNA-cleavage domain joined to a DNA-modification domain and one or two DNA sequence-specificity domains forming the C-terminus, or present as a separate subunit. When these enzymes bind to their substrates, they switch into either restriction mode to cleave the DNA, or modification mode to methylate it.

3. Type III

These enzymes are also large combination restriction-and-modification enzymes. They cleave outside of their recognition sequences and require two such sequences in opposite orientations within the same DNA molecule to accomplish cleavage; they rarely give complete digests.

❑❑❑

10

VECTOR

10.1 Plasmids

First plasmid described was found in Japan in *Shigella* species during an outbreak of dysentary in the early 1940's. Three plasmid types are the most studied: F factor (fertility factor), R plasmids (antibiotic resistance) and Col (colicinogenic).

R plasmids – some medical importance of R plasmids are as given below.

a) Penicillin was introduced for general use in early 1940's
b) 1946 - 14% of *Staphylococcus aureus* were penicillin resistant
c) 1947 - 38% penicillin resistant
d) 1969 - 59% penicillin resistant
e) 1970's - almost 100% of *S. aureus* were penicillin resistant

Col plasmids

Produce colicins, a type of bacteriocin that affect sensitive cells (Col^-) and inhibit growth. It may be self-transmissible or not ColE1 is mobilizable but it is nonconjugative

Four stages of transfers for CoI plasmids are given as below.

1) Effective contact
2) Mobilization - preparation for DNA transfer

3) DNA transfer
4) Formation of F in recipient

Four groups of plasmids based on transfer properties are as follows.

- Nontransmissible - can neither initiate contact with recipient nor transfer DNA
- Conjugative - can initiate contact with recipient bacterium
- Mobilizable - can prepare its DNA for transfer
- Self-transmissible - is both conjugative and mobilizable

Replication regulated by an RNA

RNA II (plasmid-encoded) binds to the plasmid DNA at the *ori* and then must be processed (cleaved) by RNase H to generate a free 3'-OH that is used as the primer for DNA replication. RNA I (plasmid-encoded) is complementary to RNA II and can bind to RNA II to prevent proper processing of RNA II and therefore prevent replication initiation. Since RNA I is plasmid-encoded, more RNA I will be synthesized when the plasmid concentration is high. A protein called Rop facilitates the pairing between RNA I and RNA II so that processing of RNA II is inhibited even at relatively low RNA I concentrations.

Replication regulated by RNA and protein

RepA protein (plasmid-encoded) is required for replication initiation. repA gene has two promoters, PcopB and PrepA. Transcription from PcopB will produce a mRNA with both *copB* and *repA*, and the CopB protein will repress expression from PrepA. Transcription from PrepA will produce a mRNA with only *repA* on it. When the plasmid first enters the cell, there is a burst of transcription from PrepA and no CopB protein to repress PrepA so the plasmid replicates until it reaches its copy number. Then PrepA is repressed by CopB and *repA* can only be transcribed from PcopB.

Once the plasmid has reached its copy number, synthesis of RepA is regulated by an RNA called CopA. The CopA RNA is complementary to the region of the PcopB transcript responsible for *repA* translation initiation so will bind to the PcopB mRNA. Then this double-stranded RNA is cleaved by RNase III and *repA* can't be translated. The higher the plasmid concentration, the more CopA is synthesized and the less RepA is made, thus maintaining plasmid copy number.

Replication regulated by a protein

RepA protein is an activator of replication. But RepA can also repress its own expression, so at high plasmid concentrations, RepA will repress its transcription. Here the replication also depends on the presence of iteron sequences located in the *ori*. In the handcuffing model, the iteron sequences of different plasmids can associate, thus joining the plasmids. RepA can link two plasmids by binding to their iteron sequences and preventing replication. At low plasmid concentrations, RepA binds to the *ori* and replication occurs. But at high plasmid concentrations, RepA binds to the *ori* (iteron sequences) of two plasmids and replication is inhibited.

Plasmid amplification

Chloramphenicol treatment inhibits protein synthesis; will inhibit chromosomal but not plasmid replication. Chromosomal replication requires new protein synthesis but plasmid replication uses only stable bacterial replication proteins, so plasmid replication can occur in a chemical like chloramphenicol that inhibits protein synthesis. Plasmids replicated to high copy number because repressor can't be made so can't control copy number.

Examples of Viruses

1. *Tobacco mosaic virus(TMV):* It consists (Fig. 10.1) of a single long RNA molecule enclosed in a cylindrical protein coat. It is approximately 300 nanometers long and 15 nanometers wide.

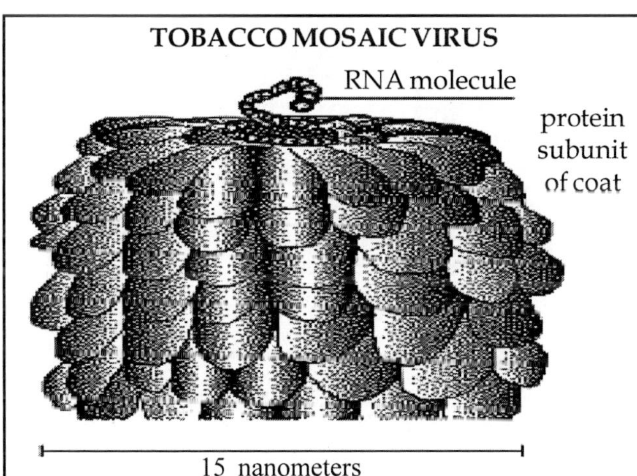

Fig. 10.1 The tobacco mosaic virus (TMV)

2. *T4 bacteriophage:* It consists (Fig. 10.2) of a DNA molecule tightly packaged in a protein head. The virus attaches to the surface of a bacteria cell and injects its DNA into the cell through its tail.

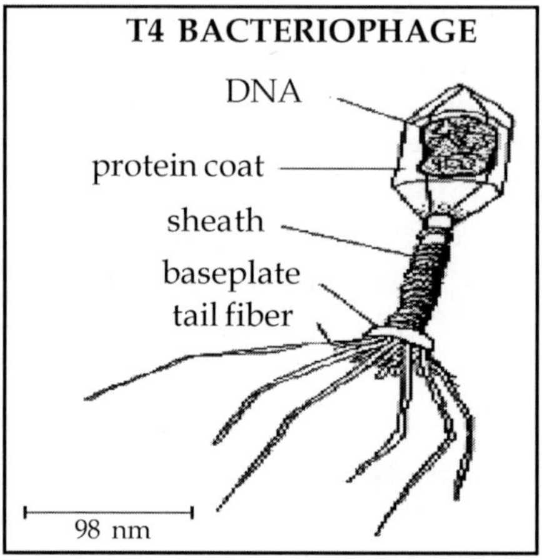

Fig. 10.2 The T4 bacteriophage

3. *Human Immunodeficiency Virus (HIV):* It consists (Fig. 10.3) of an outer lipid envelope derived from the host cell. It is a retrovirus, i.e., an RNA-containing virus that replicates by making a DNA intermediate.

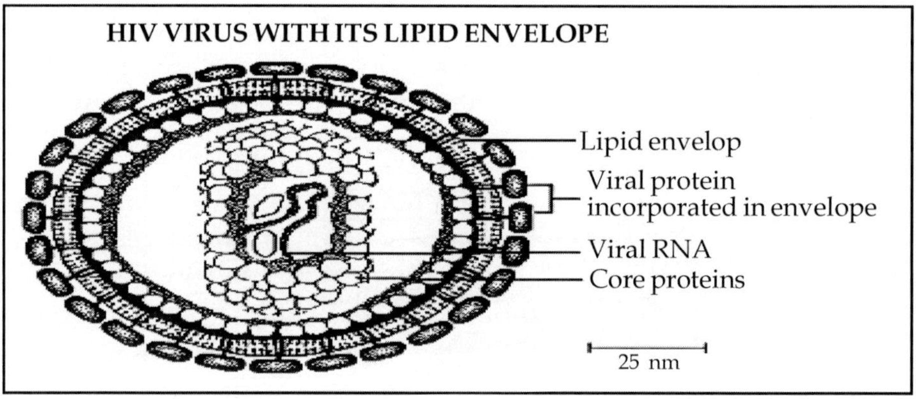

Fig. 10.3 The Human Immunodeficiency Virus (HIV).

Influenza A Virus Replication

The influenza virus is a class of viruses containing RNA as its hereditary material (Fig. 10.4).

a) It replicates by entering a host cell and using this cell's resources to produce hundreds of copies of the viral RNA.

b) The virus attaches to the outside of the host cell and its RNA enters into the cell.

c) The viral genes are transcribed and translated by the cell's enzymes and ribosomes. In this way, the virus takes over the cell's productivity.

d) Now, instead of producing only new cellular material, the cell produces hundreds of new virus particles. The new virus particles are eventually released from the cell and drift off, and some may land on a host cell of their own to pirate.

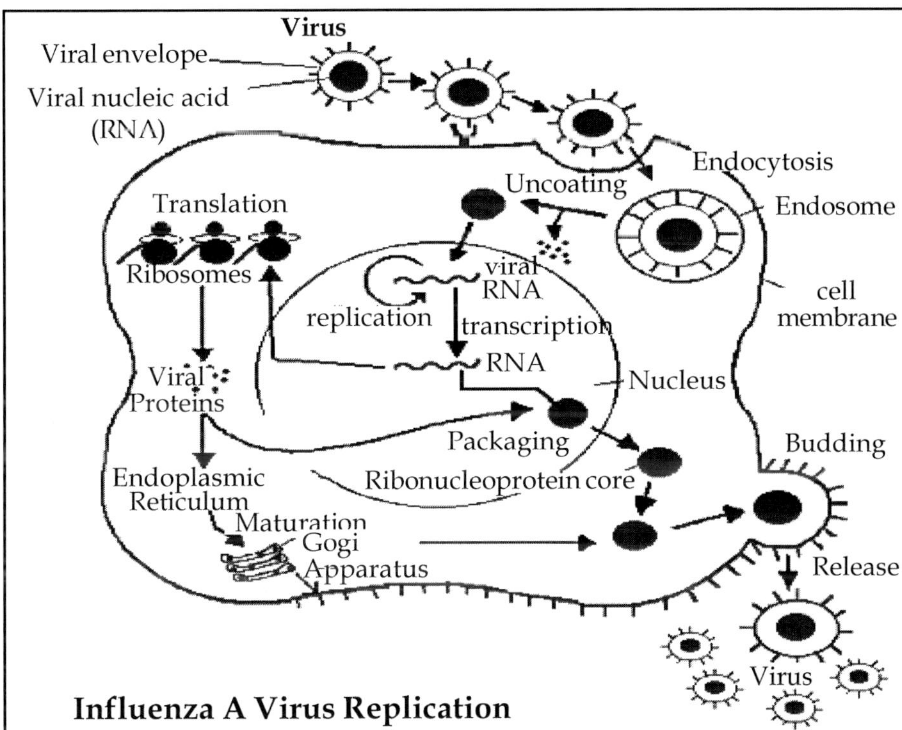

Fig. 10.4 A virus enters a host cell and infects it by reproducing its own genetic material and assembling into virus particles.

10.2 Retrovirus

Retroviruses are infectious particles consisting of an RNA genome packaged in a protein capsid, surrounded by a lipid envelope. This lipid envelope contains polypeptide chains including receptor binding proteins which link to the membrane receptors of the host cell, initiating the process of infection.

Retroviruses (Fig. 10.5) contain RNA as the hereditary material in place of the more common DNA. In addition to RNA, retrovirus particles also contain the enzyme reverse transcriptase, which causes synthesis of a complementary DNA molecule (cDNA) using virus RNA as a template.

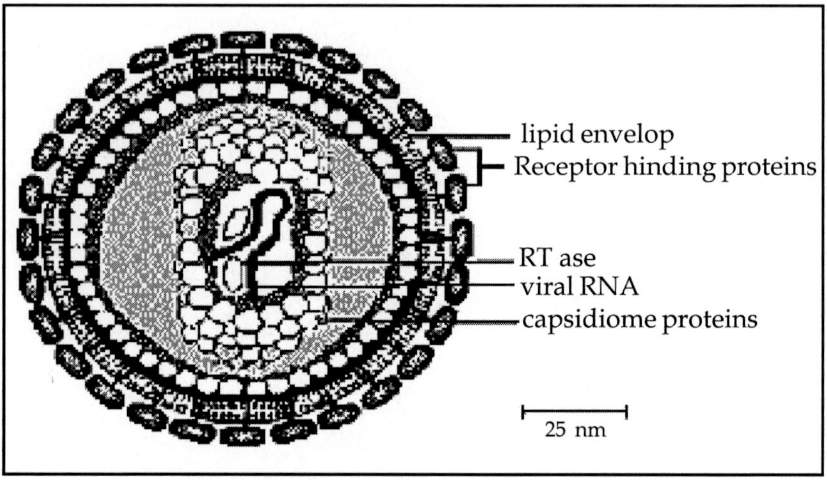

Fig. 10.5 The structure of a retrovirus

When a retrovirus infects a cell, it injects its RNA into the cytoplasm of that cell along with the reverse transcriptase enzyme. The cDNA produced from the RNA template contains the virally derived genetic instructions and allows infection of the host cell to proceed.

Replication of Retrovirus

Retroviruses contain viral RNA and several copies of reverse transcriptase (DNA polymerase). After infecting a cell, the reverse transcriptase is used to make the initial copies of viral DNA from viral RNA (Fig. 10.6). Once a DNA strand has been synthesized, a complementary viral DNA strand is made. These double strand copies of viral DNA are inserted into the host-cell chromosome and host-cell RNA polymerase is used to make virus-related RNA. These RNA strands serve as templates for making new copies of the viral chromosomal RNA and serve also as mRNA. mRNA is translated into viral proteins that are used to make the virus envelope. New viral particles are assembled, bud from the plasma membrane, and are released. An example of this process is illustrated in the replication of the retrovirus, HIV (human immunodeficiency virus).

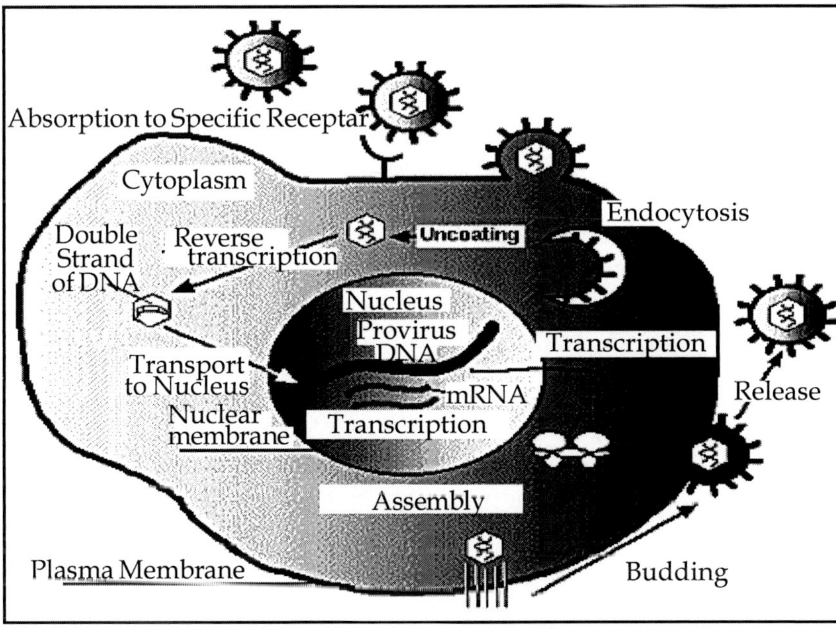

Fig. 10.6 Process by which retroviruses use a host cell to replicate.

The Life Cycle of a Retrovirus

The life cycle of a retrovirus (Fig. 10.7) includes reverse transcription and integration into the host genome. The retrovirus

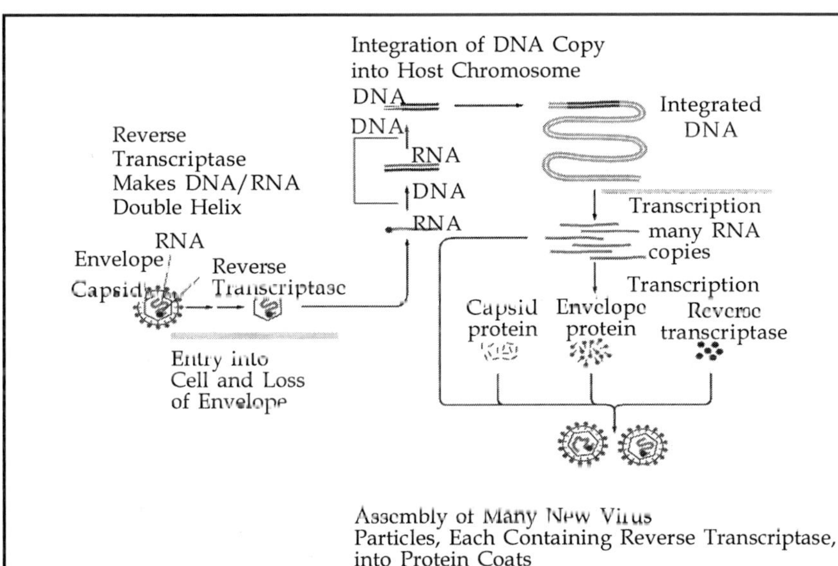

Fig. 10.7 Life Cycle of a Retrovirus

genome consists of an RNA molecule of about 8500 nucleotides; two such molecules are packaged into each viral particle. The enzyme reverse transcriptase first makes a DNA copy of the viral RNA molecule and then a second DNA strand, generating a double-stranded DNA copy of the RNA genome. The integration of this DNA double-helix into the host chromosome is then catalyzed by a virus-encoded integrase enzyme. This integration is required for the synthesis of new viral RNA molecules by the host cell RNA polymerase, the enzyme that transcribes DNA into RNA.

❑❑❑

11

PROTOCOLS

11.1 Total DNA extraction by CTAB method

S. No.	Salt	1.0 liter	Final concentration
1	NaCl	81.8 g	1.4 M
2	1 M Tris Hcl (pH 8.0)	100 ml	100 mM
3	0.5 M EDTA (disodium)	40 ml	20 mm
4	CTAB	20 g	2.0%
5	Sodium bisulphate	5 g	0.5%
6	Merceptoethanol	10 g	1.0%
7	Double distilled water	1 Lit	

Reagents

Tris (Tris hydroxyl methyl amino methane) Mol wt. 121.1

NaCl (Sodium chloride) Mol wt. 58.43

EDTA (Ethylenediamine tetraacetic acid) Mol wt. 372.2

CTAB Hexadecyl trimethyl ammonium bromide

Molar solution means Molecular weight equivalent dissolved to make one litre solution.

Precautions

1. Handle liquid nitrogen carefully.
2. Always wear gloves while DNA extraction because our skin surface has high DNAse activity.

3. Always number the tubes while DNA extraction.
4. Alcohol can erase most forms of marker inks. Be careful not to spill alcohol on marked parts of the tubes.

Preparation reagents

1M Tris (pH 8.0)	Add 121.1 g Tris in about 800 ml water. Stir vigorously, using a magnetic stirrer. Adjust pH to 8.0 by adding 42 ml conc. HCl. Cool the solution to room temperature and adjust volume of the solution to 1.0 litre with ddH_2O. Sterilize the solution by autoclaving. *(IMP Note: If solution turns yellow, discard it).*
0.5M EDTA (pH 8.0)	Add 186.1 g $Na_2EDTA.2H_2O$ in 800 ml ddH_2O. Stir vigorously, using a magnetic stirrer. Add about 20g NaCl pellets (EDTA will not dissolve without adding NACl). Adjust volume of the solution to 1.0 litre with ddH_2O. Sterilize the solution by autoclaving.
2.0% CTAB	It is difficult to dissolve. Add 2.0 g CTAB in about 900 ml sterile ddH_2O and stir on a hot magnetic plate.
2 X CTAB Buffer	Mix the components 1 to 4 of table 1 in the given order. This can be stored at room temperature. Sodium bisulphate and merceptoethanol are added just prior to using the buffer.
TE Buffer PH 8.0	10mM Tris HCl, pH 8.0 + 1mM EDTA, pH 8.0 (For 100 ml TE buffer, mix 1 ml of 1M Tris, 0.2 ml of 0.5M EDTA and 98.8 ml ddH_2O

Extraction

- Collect the Sample.
- Homogenize about 10 g of leaf sample in liquid nitrogen and transfer the powder in a tube.

- Now add 15 ml of 2X CTAB buffer to the tube. Re-suspend leaves in buffer by inverting the tubes and incubate at 65⁰C for 30 min in a water bath. Mix the samples occasionally while maintaining at 65⁰C.
- *Now add 15ml of chloroform :* isoamyl alcohol (24:1) and swirl the tubes until it make emulsion. The tubes are placed on a rotary shaker for 15min to 60min at room temperature.
- Centrifuge it for 15 min at 5000g at room temperature.
- Transfer supernatant to a clean sterile tube and add 5µ Rnase (10mg/ml). Incubate at 37⁰C in a precision oven for 60 min.
- *Repeat chloroform :* isoamyl alcohol (24:1) extraction. Gently shake for 15min on a rotary shaker and centrifuge for 15min. Transfer supernatant.
- Add about 0.8 volume of isopropyl alcohol (cooled at 4⁰C). Invert the tubes gently several times. The DNA will form white cotton like precipitate and will float atop. *(Note: Good quality DNA will float atop while poor quality DNA will settle down).*
- Transfer the DNA in to a microcentrifuge tube and wash the DNA 2 - 3 times with 70% ethanol.
- Dry the ethanol completely by turning down microfuge tubes on a blotting paper.
- Add 200 - 1000µ of TE (Tris EDTA pH 8.0) depending upon the quantity of DNA. Leave the tube for few hours at room temperature. If the pellet does not dissolve, either leave the samples overnight at room temperature or heat the samples at 60⁰C in a water bath for 10 - 15 min.
- DNA can be stored at 4⁰C for several months.

11.2 Extraction, Purification Estimation of DNA

The aim of genomic DNA preparation technique is to isolate high molecular weight DNA of sufficient purity. There are two factors affect for the size of DNA isolated, (1) Shear forces and (2) Nuclease activity Lysates should be treated gently to minimize shear forces and for reduce nuclease activity tissue should be frozen quickly and thawed only in the presence of extraction buffer that contains detergent and a high concentration of EDTA.

Preparation

The nonionic detergent cetyltrimethyl ammonium bromide (CTAB) is used to librates and complex with total cellular nucleic acids. This general procedure has been used on a wide array of plant genera and tissue types.

Reagents

1. liquid Nitrogen
2. Isolation buffer 2% CTAB, 1.4M NaCl, 20mMEDTA, 0.1M Tris-HCl, pH 8.0
3. 2-Mercaptoethanol
4. Chloroform:isoamyl alcohol (24:1)
5. RNase: Dissolve RNase at a conc. Of 10mg/ml in 10mM Tris-HCl (pH 7.5) and 15mM NaCl. Heat to 100^{0}C for 15 min. and allow cooling slowly to room temperature. Dispense into aliquots and store at – 20^{0}C.
6. Equilibrated phenol or Phenol:Chloroform (1:1).
7. *TE Buffer:* 10mM Tris-HCl, 1mM EDTA, pH 8.0
8. 3M Sodium acetate
9. 100% Ethanol
10. Wash solution (a) 76% Ethanol, 0.2M Sodium acetate
    (b) 76% Ethanol, 0.2M Ammonium acetate

Method

- Add 2- Mercaptoethanol to the required amount of CTAB extraction solution to give a final concentration of 2% (V/V), heat this solution to 65^{0}C.
- Chill a mortar and pestle with liquid nitrogen.Grind approximately 1g of fresh leaf tissue to a fine powder and transfer the frozen tissue into a 50ml polypropylene centrifuge tube. (*Note*: Use young tissue and avoid larger stems and veins to achieve the highest DNA yield with minimal polysaccharide contamination)
- Add 15ml of preheated 2- Mercaptoethanol / CTAB extraction solution to the pulverized tissue and mix to wet thoroughly.

- Incubate for 60-90 min. with occasional mixing. (A 60 min. incubation results in larger DNA yield. If maximum yield is not important than keep only incubation for 15min. If the tissue contains larger amounts of phenolic compounds, 1% (v/v) poluvinylpyrrolidone may be added to absorb them).
- Extract the homogenate with an equal volume of 24:1 chloroform: isoamyl alcohol and mix well.
- Centrifuge for 10min at 8000 rpm at room temperature.
- Transfer the aqueous phase to another 50ml polypropelene centrifuge tube. Add heat treated RNAse. A to a final concentration of 100g/ml. Mix, and incubate at room temperature for 30min.
- Add 0.6 volume of ice-cold isopropanol.
- Spool the precipitate out of the solution and purify it.

Purification

(A) Phenol extraction

1. Keep DNA in eppendrof tube containing 1ml of TE. Dissolve DNA at 4^0C overnight. If the DNA is needed immediately, it may be incubated at 37^0C for one hour.
2. Add equilibrated phenol: chloroform (1:1) equal to the total amount of TE added as above step 1. Mix thoroughly for 10 min and centrifuge it at room temperature. Upon centrifugation, the phenol layer, which contains some proteins and lipids, moves toward the bottom of the tube, while the aqueous layer containing the DNA remains as top. The denatured protein forms an insoluble layer at the interphase.
3. Transfer the aqueous layer to a new tube and extract with an eual volume of chloroform. Mix well, centrifuge for 5 min. Transfer the supernatant to a new tube.

(B) Ethanol extraction

1. Keep DNA in eppendrof tube containing 1ml of TE. Keep DNA at 4^0C overnight.
2. Add $1/10^{th}$ volume of 3M sodium acetate, pH 5.5 and double volume of ethanol and mix well.

(C) DNA wash

- After the phenol and or ethanol extraction/precipitation wash the DNA with wash solution 1 (1 to 2 ml), leave it for 30 min.
- Rinse DNA with wash solution 2 (1 to 2 ml), and dry it.
- Transfer DNA to eppendrof tube containing 0.5 1.0ml of TE, and dissolve the DNA. Store at 4^{0}C.

11.3 Potassium Acetate Procedure

This method represents a reasonable alternative to the CTAB procedures. Its key step relies on the simultaneous precipitation of proteins as well as polysaccharides by high concentration of potassium acetate in the presence of SDS. Organic extractions are not required in this method is the advantage of this procedure.

Reagents

a) Liquid nitrogen

b) Isolation buffer (100mM Tris HCl, pH 8.0, 50mM EDTA, 500mM NaCl, 10mM 2-mercaptoethanol, 1.25% SDS (w/v).

c) 5M potassium acetate

d) Isopropanol

e) *TE buffer:* 10mM Tris HCl, 1mM EDTA, pH 8.0.

f) *RNAse:* Dissolve RNAse 10mg/ml in 10mM Tris-HCl, pH 7.5, and 15mM NaCl. Heat at 100^{0}C for 10-15 min and allow cooling at room temperature. Dispense into aliquots and store at - 20^{0}C.

g) 3M sodium acetate, pH 8.0

h) Ehanol 100%

i) Ethanol 70%

Procedure

1) Homogenize 3-5g plant sample using liquid nitrogen.

2) Transfer the powder as fast as possible into 15ml of pre-warmed (65^{0}C) solation buffer in a polypropylene tube.

3) Incubate it for 15 - 20 min at 65^{0}C in a water bath.

4) Add 5ml of potassium acetate and mix well, incubate on ice for 30min. Here the proteins as well as polysaccharides are precipitated together with the insoluble potassium dodecyl sulfate.
5) Centrifuge it at 4^{0}C for 20min. Precipitate DNA by the addition of 1.0ml of ice cold 100% isopropanol.
6) Incubate it at - 20^{0}C overnight
7) Wash the DNA pellet with 70% ethanol. Re-suspend DNA pellet in 5 ml of TE buffer.
8) Add 1 ml of RNAse (10mg/ml). Incubate at 37^{0}C for 15 min.
9) Add 1/10th volume of 3M sodium acetate and 2.5 volume of 100% ethanol.
10) Mix well and incubate at - 20^{0}C for overnight.
11) Centrifuge at 20,000g for 10 min and wash DNA pellet with 70% Ethanol, and recentrifuge it.
12) Remove the supernatant.
13) Resuspended the DNA pellet in 0.5 - 1.0ml of TE buffer and transfer to 1.5ml eppendrof tube.
14) Store DNA at 4^{0}C. Use the DNA for PCR analysis.

11.4.1 DNA Estimation

DNA quantification may be achieved by analysis on an agarose electrophoresis gel with ethidium bromide staining.

Reagents

1) Ethidium bromide 10mg/ml in distilled water.
2) 0.8% Agarose in electrophoresis buffer
3) Electrophoresis buffer: TAE or TBE
4) 10X loading Dye: 40% bromophenol blue, 40% xylene Cyanol FF, 50% Glycerol.

Procedure

- Load about 2 µg of undigested DNA on the gel along with 0.5 µg of molecular weight standard (lambda DNA cu with Hindiii).
- Visualize the DNA on a gel following electrophoresis and after staining with ethidium bromide.

- Compare the migration of the sample with the standard.
- The bands for the samples must appear as a distinct solid fluorescence. (Diffused bands reaching toward lower molecular weight indicates degradation of DNA fragments.

DNA Quality

1) Keep 5.0 µl of the DNA sample in the cuvette. Add 95 µl of TE. Mix well.
2) Read the O.D. at 280 nm.
3) Read the O.D. at 280 nm.
4) Calculate the ratio of O.D. 260 to 280. a ratio value of 1.8 suggests a highly pure preparations of DNA. If the ratio is 1.8 - 2.0, the absorption is probably due to nucleic acids. A ratio less than 1.8 indicates that there may be proteins and / or other UV absorbers in the sample, in which cases it is advisible to reprecipitate the DNA. A ratio higher than 2.0 indicates the samples may be contaminated with chloroform or phenol and should be reprecipitated with ethanol extraction procedure. Repeat the O.D. values for the other samples at 260 and 280nm by repeating steps described as above.

11.5 Restriction Enzymes

Recombinant DNA technology or "gene splicing," as it is frequently called, often involves insertion of selected DNA sequences from a variety of sources into plasmids, which are then transformed into bacteria. The resultant recombinant bacterial strain may now express the plasmid genes as well as whatever foreign gene(s) that was inserted.

This process would not be possible without accurate and predictable ways of cutting the DNA of the plasmid and the foreign source. Restriction endonuclease enzymes provide this service. Restriction endonucleases, often called restriction enzymes, cut DNA within the molecule by hydrolyzing the phosphodiester bonds between the nucleotides. Restriction enzymes were discovered in bacteria and there are now more than 1200 known restriction enzyme types. These enzymes are named using a simple system. EcoRI, for example, was isolated from E. coli and was the first enzyme isolated from a particular strain, hence the designation of I. HaeIII was isolated from

Haemophilus aegyptius. HindIII was isolated from Haemophilus influenzae, and was the third enzyme discovered in a particular strain.

Usually, restriction enzymes only cut the DNA at or near a very specific nucleotide sequence known as a recognition site. This "restrictive" nature of these enzymes allows molecular biologists to pin-point exactly where the DNA is to be cut. In many cases, the recognition site of a restriction enzyme consists of a palindromic sequence where the order of the bases is read the same on each side of the helix. The two sides of the helix are antiparallel and the recognition sequence reads from 5′ to 3′ on either side of the helix. For example the recognition site for EcoRI is as follows;

GAATTC

CTTAAG

Some restriction enzymes cut both sides of the helix at the same point. HaeIII is such an enzyme and the fragments produced are "blunt" ended. Many restriction enzymes make asymetric cuts in the helix. EcoRI cuts DNA at the recognition site in the following fashion;

G?AATTC

CTTAA?G.

This type of asymmetric cut leaves a single stranded sequence hanging on either 5′ or 3′ end. Such a trailing end is often called a "sticky" end since it can bond by hydrogen bonding to a complementary single stranded sequence. The sticky ends make it rather easy to insert DNA into a plasmid. If the plasmid and DNA to be inserted are cut with the same restriction enzyme, the sticky ends will match and will anneal using DNA ligase enzyme. The result is a recombinant plasmid with a foreign fragment incorporated. Cleavage of DNA by a restriction enzyme is called digestion and is usually carried out at 37°C in a small volume (e.g., 20 μl) of buffer and salt. Although restriction enzymes are active over a fairly broad range of conditions, optimal conditions have been determined for each restriction enzyme using a buffer which typically contains 10-100mM Tris-HCl pH 8.0, a salt (usually 0-150 mM NaCl), and 10 mM of Magnesium Chloride. Restriction enzyme digestion involves adding enough restriction enzyme to the substrate DNA in the appropriate buffer and letting the digest incubate at the appropriate temperature for at least one hour. Restriction enzymes are measured functionally in units (μ), which are defined as the amount of restriction enzyme that will completely digest a defined substrate

DNA in a defined time at a specific temperature. When stored at the recommended conditions (-20°C in storage buffer containing 50% glycerol), restriction enzymes can remain active for many months. However, once they are diluted into digestion buffer and incubated at 37°C, they lose activity within hours.

Procedure

1. EcoRI digests

A. Restriction Enzyme Digestion of pUC19 DNA

(Note- All components involved in digestion should be kept on ice)

1. Prepare an EcoRI digest of pUC19 DNA by adding the required components to a clean microtube in the following order:

   14μl distilled water

   2μl 10 X EcoRI Buffer

   3μl of pUC19 DNA (uncut)

   lμl of Restriction enzyme EcoRI

   Place your initials on this tube and label it pUC19/EcoRI

2. Flick the tube to mix well and spin for 5 sec in the microfuge to bring all the components to the bottom.

3. Incubate this digest mixture at 37°C for 20 minutes (in a 37°C waterbath).

B. Restriction Enzyme Digestion of pRAS2 DNA

(Note- All components must be kept on ice)

1. Prepare an EcoRI digest of pRAS2 DNA by adding the components to a clean microfuge tube in the following order:

   14μl-distilled water

   2μl 10 X EcoRI buffer

   3μl pRAS2 DNA (uncut)

   lμl Restriction enzyme EcoRI

   Place your initials on this tube and label it pRAS2/EcoRI

2. Flick the tube and spin for 5 seconds in the microfuge to bring all the components to the bottom.

3. Incubate at 37°C in an incubator or waterbath for 20 minutes.

C. Restriction Enzyme Digestion of Genomic Wheat Germ DNA using EcoRI (optional)

(Note- Keep all components on ice)

1. Prepare an EcoRI digest of genomic Wheat Germ DNA by adding the components to a clean microtube in the following order:

   30µl distilled water
   5µl 10 X EcoRI Buffer
   10µl Wheat germ DNA (prepared in earlier exercise)
   5µl Restriction Enzyme EcoRI

   Place your initials on this tube and label it genomic/EcoRI

2. Flick the tube and spin it for 5 seconds in the microfuge to bring all components to the bottom.

3. Incubate this digest mixture in a 37°C waterbath for several hours. (note- the instructor may have to complete this and prepare a sample for electrophoresis to be done the next day)

D. Restriction Enzyme Digestion of Lambda phage DNA using EcoRI

(Note - All components must be kept on ice)

1. Prepare an EcoRI digest of Lambda DNA by adding the components to a clean microtube in the following order:

   14µl distilled water
   2µl 10 X EcoRI Buffer
   3µl Lambda DNA (uncut)
   1µl Restriction enzyme EcoRI

   Place your initials on this tube and label it Lambda/EcoRI

2. Flick the tube and spin for 5 second in the microfuge to bring all the components to the bottom.

3. Incubate this digest mixture at 37°C (waterbath) for 20 minutes.

E. Preparation of samples for electrophoresis

1. When the incubation of the pUC19 digest ends, transfer this tube to ice for 5 minutes.

2. Add 2µl of loading dye-Label this tube pUC19/EcoRI. Place this tube on ice.

3. When the incubation of the pRAS2 digest ends, add 2μl of loading dye directly to this digest- flick the tube to mix and microfuge for a few seconds. Transfer this tube to ice.
4. When the incubation of the genomic Wheat Germ DNA digest ends, transfer this tube to ice for 5 minutes.
5. Add 2μl of loading dye. Flick this tube and briefly spin and label it Genomic/EcoRI. Place this tube on ice.
6. When the incubation of the Lambda DNA digest ends, transfer this tube to ice for 5 minutes.
7. Add 2μl of loading dye. Label this tube Lambda/EcoRI and place on ice.

Note: At this point each group should have 4 tubes (4 digests), which contain loading dye. These tubes can be stored on ice overnight.

F. Electrophoresis of EcoRI digests

1. Prepare a 1% Agarose Gel (eight well comb) and allow it to stand for 20 minutes
2. After removing the dams, add the chamber buffer. Then, after a few minutes to allow the gel to soften, remove the comb by pulling straight up.
   (Note-The instructor can save time by doing steps 1 & 2 prior to class)
3. Load 20μl of each sample (the one kilobase ladder and 4 digests) into the wells in the order indicated.

   Lane 1 - one kb ladder
   Lane 2 - leave empty or use for practice loading
   Lane 3 - genomic Wheat Germ digest
   Lane 4 - pUC19 digest
   Lane 5 - pRAS2 digest
   Lane 6 - lambda digest
   Lane 7- lambda digest (HindIII) if available
4. After the samples have been loaded, close the lid, connect the power supply, and switch on. Set the voltage at 110 volts.
5. Electrophorese for about 30 minutes until the dye has reached midway between the second and third red stripes. Switch off the power supply and open the lid.

6. Wearing gloves, remove the gel and stain with methylene blue in a tray for 30 minutes and move the gel to a tray of distilled water for 15 minutes to destain.
7. After staining, place the gel in a tray (half of an empty tip box) using a spatula and bring to the light box.
8. The instructor will photograph your gel. (You may destain overnight if needed.)
9. After photographing, discard the gel and gloves in the waste container for proper disposal.

Preparation instructions

1. There are four digests- pUC19, pRAS2, lambda phage, and genomic. All digests should be prepared, incubated, prepared for electrophoresis and store.
2. The gels should be made the second day and may be poured ahead of time and stored under chamber buffer.
3. The gels should be stained and photographed right after being run since the bands become diffuse with time.
4. As always, familiar with all transfer techniques, electrophoresis procedures and care and cleaning of equipment. Stress safety when dealing with high voltage.

Materials List for Restriction Enzyme Exercise

1. restriction enzyme EcoRI for plasmid digests 3µl (1µl/digest)
2. restriction Enzyme EcoRI for genomic digest 5µl (optional)
3. 10 X buffer for plasmid digests using EcoRI, 6µl
4. 10 X buffer for genomic digest using EcoRI, 5µl (optional)
5. pUC19 DNA uncut, 3µl
6. pRAS2 DNA uncut, 3µl
7. lambda DNA uncut, 3µl
8. microtubes, 0.65 ml clear (as needed)
9. microfuge, 10k with 2 rotors
10. ice

11. float for water bath
12. water Bath, 37 C
13. loading dye - as needed
14. p20 and tips
15. marking pen
16. one kilobase ladder, 20μl
17. TE Buffer, 0.1 ml
18. electrophoresis chamber, gel casting tray, combs, dams, power supply and cords
19. 1% Agarose – 25 ml
20. gloves
21. methylene blue solution
22. light table and photodyne camera
23. staining and destaining trays

❑❑❑

12

Chemicals Reagents Properties

Acetic acid

Molecular weight	60.05
Melting point (^{0}C)	17
Boiling point (^{0}C)	118
Flash point (^{0}C)	40
Auto-ignition point (^{0}C)	426
Specific gravity at 20^{0}C	1.05
Molarity of the concentrated acid (M)	17.4
1 M solution	57.5 ml concentrated acid per litre

Acetone

Molecular weight	58.1
Melting point (^{0}C)	- 95
Boiling point (^{0}C)	56
Flash point (^{0}C)	- 20
Specific gravity at 20^{0}C	0.79

Acrylamide

Molecular weight	71.1

Actinomycin D

Molecular weight	1255.5
Solubility in water	Slightly soluble
Mode of action	Inhibits transcription by binding to dG (3′-5′)dC
Note	*Store dry at – 20°C, in the dark*

Adenine

Molecular weight	135.1
Solubility (25°C)	90 mg/100 ml water
pK_{a1}	< 1.0
pK_{a2}	4.1
pK_{a3}	9.8

Adenosine

Molecular weight	267.2
Solubility in water	Soluble
pK_{a1}	3.5
pK_{a2}	12.5

SAM (S-Adenosylmethionine)

Molecular weight	399.4
Solubility in water	Very soluble

ADP (Adenosine diphosphate)

Molecular weight	427.2
Solubility in water	Very soluble
pK_{a1}	3.8
pK_{a2}	6.1 - 6.7

AMP (Adenosine monophosphate)

Molecular weight	347.2
Solubility in water	Very soluble
pK_{a1}	3.7
pK_{a2}	5.9

Ammonium persulfate

Molecular weight	228.2
Solubility (0^0C)	58.2g/100 ml water

ATP (Adenosine triphosphate)

Molecular weight	507.2
Solubility in water	Very soluble
pK_{a1}	4.1
pK_{a2}	6.0 - 7.0

Bromocresol green (3′3″5′5″ Tetrabromo – m-cresolsulfonephthalein)

Molecular weight	698.0
Use	As Gel electrophoresis marker dye

Bromophenol blue (3′3″5′5″ Tetrabromo phenolsulfonephthalein)

Molecular weight	670.0
Use	As Gel electrophoresis marker dye

n-Butanol

Molecular weight	74.1
Melting point (^0C)	-89
Boiling point (^0C)	118
Flash point (^0C)	29
Specific gravity	0.81

cAMP (3′ 5′ Cyclic adenosine monophosphate)

Molecular weight	329.2
Solubility in water	Very soluble

Chloroform

Molecular weight	119.4
Melting point (^{0}C)	- 63
Boiling point (^{0}C)	61
Specific gravity	1.48

Coenzyme A

Molecular weight	767.6
Solubility in water	Very soluble

Coomassie blue

Molecular weight	826.0
Use	Protein stain in gel electrophoresis

CTAB (Cetyltrimethylammonium bromide)

Molecular weight	364.5
Critical micelle concentration mM	1.0
Micellar weight	62000

Cytidine

Molecular weight	243.2
Solubility in water	Very soluble
pK_{a1}	4.2
pK_{a2}	12.3

Cytosine

Molecular weight	111.1
Solubility in water	77mg/100ml water
pK_{a1}	4.4
pK_{a2}	12.2

dADP (2′ Deoxyadenosine 5′ diphosphate)

Molecular weight	411.2
Solubility in water	Very soluble

damp (2′ Deoxyadenosine 5′ monophosphate)

Molecular weight	331.2
Solubility in water	Very soluble
pK_{a1}	4.4
pK_{a2}	6.4

dATP (2′ Deoxyadenosine 5′ triphosphate)

Molecular weight	491.2
Solubility in water	Very soluble
pK_{a1}	4.8
pK_{a2}	6.8

dCDP (2′ Deoxycytidine 5′ diphosphate)

Molecular weight	387.2
Solubility in water	Very soluble

dCMP (2′ Deoxycytidine 5′ Monophosphate)

Molecular weight	307.2
Solubility in water	Very soluble
pK_{a1}	4.6
pK_{a2}	6.6
pK_{a3}	13.2

dCTP(2′ Deoxycytidine 5′ phosphate)

Molecular weight	467.2
Solubility in water	Very soluble

ddATP(2′3′ Dideoxyadenosine 5′ phosphate)

Molecular weight	457.2
Solubility in water	Very soluble

ddCTP(2′3′ Dideoxycytidine 5′ phosphate)

Molecular weight	451.2
Solubility in water	Very soluble

ddGTP (2′3′ Dideoxyguanosine 5′ phosphate)

Molecular weight	491.2
Solubility in water	Very soluble

ddTTP (2′3′ Dideoxythymidine 5′ phosphate)

Molecular weight	466.2
Solubility in water	Very soluble

2′ Deoxyadenosine

Molecular weight	251.22
Solubility in water	soluble
pK_a	3.8

2′ Deoxycytidine

Molecular weight	227.2
pK_{a1}	4.2
pK_{a2}	>13.0

2′ Deoxyguanosine

Molecular weight	267.2
Solubility in water	soluble
pK_a	2.5

2′ Deoxythymidine

Molecular weight	242.2
pK_{a1}	9.8
pK_{a2}	>13.0

dGDP (2′ Deoxyguanosine 5′ diphosphate)

Molecular weight	427.2
Solubility in water	Very soluble

dGMP (2′ Deoxyguanosine 5′ monophosphate)

Molecular weight	347.2
Solubility in water	Very soluble
pK_{a1}	2.9
pK_{a2}	6.4
pK_{a3}	9.7

dGTP (2′ Deoxyguanosine 5′ triphosphate)

Molecular weight	507.2
Solubility in water	Very soluble
pK_{a1}	3.5
pK_{a2}	6.5
pK_{a3}	9.7

Diethyl pyrocarbonate Rnase inhibitor

Molecular weight	162.2
Activity	Ethoxyformylates proteins

dITP (2′ Deoxyinosine 5′ triphosphate)

Molecular weight	492.2
Solubility in water	Very soluble

dTDP (2′ Deoxythymidine 5′ diphosphate)

Molecular weight	402.2
Solubility in water	Very soluble

dTMP (2′ Deoxythymidine 5′ monophosphate)

Molecular weight	322.2
Solubility in water	Very soluble
pK_{a1}	1.6
pK_{a2}	6.5
pK_{a3}	10.0

dTTP (2′ Deoxythymidine 5′ triphosphate)

Molecular weight	482.2
Solubility in water	Very soluble

dUTP (2′ Deoxyuridine 5′ triphosphate)

Molecular weight	468.2
Solubility in water	Very soluble

EDTA (Ethylenediaminetetraacetic acid)

Molecular weight	372.2
Solubility (20^0 C)	11.1g /100 ml water
pK_{a1} (25^{0}C)	1.99
pK_{a2} (25^{0}C)	2.67
pK_{a3} (25^{0}C)	6.16
pK_{a4} (25^{0}C)	10.26

GDP (Guanosine 5′ diphosphate)

Molecular weight	443.2
Solubility (20^0 C)	Very soluble
pK_{a1} (25^{0}C)	<1.0
pK_{a2} (25^{0}C)	2.9
pK_{a3} (25^{0}C)	6.3
pK_{a4} (25^{0}C)	9.6

GMP (Guanosine 5′ monophosphate)

Molecular weight	363.2
Solubility (20^0 C)	Very soluble
pK_{a1}	2.4
pK_{a2}	6.1
pK_{a3}	9.4
pK_{a4}	

GTP (Guanosine 5′ triphosphate)

Molecular weight	347.2
Solubility (20^0 C)	Very soluble
pK_{a1} (25^{0}C)	<1.0
pK_{a2} (25^{0}C)	3.3
pK_{a3} (25^{0}C)	6.3
pK_{a4} (25^{0}C)	9.3

Guanine

Molecular weight	151.1
Solubility (40^0 C)	4mg/100 ml water
pK_{a1}	3.3
pK_{a2}	9.2
pK_{a3}	12.0

Guanosine

Molecular weight	283.2
Solubility in water	Soluble
pK_{a1}	1.6
pK_{a2}	9.2
pK_{a3}	12.4

5 Hydroxymethylcytosine

Molecular weight	141.1
Solubility in water	Slightly soluble
pK_{a1}	4.3
pK_{a2}	13.0

5 Hydroxymethyl 2′deoxycytidine

Molecular weight	257.2
pK_a	3.5

Hypoxanthine

Molecular weight	136.1
Solubility (19^0 C)	70 mg/100 ml water
pK_{a1}	2.0
pK_{a2}	8.9
pK_{a3}	12.1

Inosine

Molecular weight	268.2
Solubility (19^0 C)	1.6 g/100 ml water
pK_{a1}	1.2
pK_{a2}	8.9
pK_{a3}	12.5

ITP (Inosine 5′ triphosphate)

Molecular weight	508.2
Solubility (19^0 C)	Very soluble
pK_a	6.9

5 Methylcytidine

Molecular weight	257.2
pK_{a1}	4.3
pK_{a2}	<13.0

5 Methylcytosine

Molecular weight	125.1
Solubility (25^{0}C)	450 mg/100 ml water
pK_{a1}	4.6
pK_{a2}	12.4

5 Methyl 2′ deoxycytidine

Molecular weight	241.2
Solubility in water	Soluble
pK_{a1}	4.4
pK_{a2}	>13.0

N N′ Methylenebisacrylamide

Molecular weight	154.2

7 Methylguanine

Molecular weight	165.2
pK_{a1}	<1.0
pK_{a2}	3.5
pK_{a3}	9.9

7 Methylguanosine

Molecular weight	297.3
pK_a	7.0

NAD (Nicotinamide adenine dinucleotide)

Molecular weight	663.4
Solubility in water	Very soluble

NADH (Nicotinamide adenine dinucleotide reduced)

Molecular weight	665.4
Solubility in water	Very soluble

NADP (Nicotinamide adenine dinucleotide phosphate)

Molecular weight	743.4
Solubility in water	Very soluble

NADPH (Nicotinamide adenine dinucleotide phosphate reduced)

Molecular weight	745.4
Solubility in water	Very soluble

Orotic acid

Molecular weight	156.1
Solubility (18^{0}C)	180 mg/100 ml water
pK_{a1}	2.4
pK_{a2}	9.5
pK_{a3}	13.0

Orotidine

Molecular weight	288.2
Solubility in water	Soluble

Orthophosphoricacid

Molecular weight	98.00
Melting point (^{0}C)	22
Boiling point (^{0}C)	261
Specific gravity at 20^{0}C	1.70
Molarity of the conc. acid	14.7
1M solution	67.8 ml conc.acid/litre

Pseudouridine

Molecular weight	244.2
pK_{a1}	9.0
pK_{a2}	>13.0

Sodium hydroxide

Molecular weight	40.00
Melting point (^{0}C)	318
Boiling point (^{0}C)	1390
Specific gravity at 20^{0}C	2.12

Thymine

Molecular weight	126.1
Solubility (25⁰C)	400 mg/100 ml water
pK_{a1}	- 0
pK_{a2}	9.9
pK_{a3}	>13.0

UDP (Uridine 5′ diphosphate)

Molecular weight	404.2
Solubility in water	Very soluble

UMP (Uridine 5′ monophosphate)

Molecular weight	324.2
Solubility in water	Very soluble
pK_{a1}	1.0
pK_{a2}	6.4
pK_{a3}	9.4

Uracil

Molecular weight	112.1
Solubility in water	360 mg / 100 ml water
pK_{a1}	< 0.5
pK_{a2}	9.5
pK_{a3}	>13.0

Uric acid

Molecular weight	267.2
Solubility (20⁰C)	2 mg / 100 ml water
pK_{a1}	5.4
pK_{a2}	11.3

Uridine

Molecular weight	244.2
Solubility in water	Soluble
pK_{a1}	9.2
pK_{a2}	12.5

UTP (Uridine 5′ triphosphate)

Molecular weight	484.2
Solubility in water	Very soluble
pK_{a1}	1.0
pK_{a2}	6.6
pK_{a3}	9.6

Xanthine

Molecular weight	152.1
Solubility (20^{0}C)	50 mg / 100 ml water
pK_{a1}	0.8
pK_{a2}	7.4
pK_{a3}	11.1

❑❑❑

Glossary

30 nm fiber.:A chromatin structure in which nucleosomes are coiled into a solenoid 30 nm in diameter.

A

A.F.I.S.: Automated Fingerprint Identification System

Acquired mutations: Gene changes that arise within individual cells and accumulate throughout a person's lifetime; also called somatic mutations.

Addition:A mutation caused by the insertion of at least one extra nucleotide base in a DNA sequence.

Adenine:A nitrogenous base. Pairs with thymine in DNA molecules. OR A purine base; one of the four molecules containing nitrogen present in the nucleic acids DNA and RNA; designated by letter A. OR Designated by the letter A, it is a purine base. One of the four base molecules present in DNA. OR An organic base, one of the four monomers or building blocks from which DNA is made.

AFLP(amplified fragment length polymorphism): Amplified fragment length polymorphism (AFLP) is a PCR-based technique that involves restriction of genomic DNA followed by ligation of adaptors to the fragments generated and selective PCR amplification of a subset of these fragments. The amplified fragments are separated on a sequencing gel and visualized by autoradiography or fluorescent sequencing equipment.

Agarose: One of a series of alternative forms of a gene (or VNTR) at a specific locus in a genome.OR A jello like substance derived from seaweed that is used as a medium for the size separation of DNA and proteins by the process of electrophoresis. OR An alternative form of a gene.

Alleles: Alternative forms of a genetic characteristic. OR Variant forms of the same gene. Different alleles produce variations in inherited characteristics such as eye color or blood type.

ALU-PCR :PCR using a primer that anneals to Alu repeats to amplify DNA located between two oppositely oriented Alu sequences. Used as a method of obtaining a fingerprint of bands from an uncharacterized human DNA.

Alzheimer's disease: A disease that causes memory loss, personality changes, dementia and, ultimately, death. Not all cases are inherited, but genes have been found for familial forms of alzheimer's disease.

Amelogenin : A gene present on the X and Y sex chromosomes that is used in DNA identification testing to determine the gender of the donor of the DNA in a biological sample.OR A PCR amplified locus that varies between males (212 and 218 bp) and females (212 bp only).

Amino acid : Small molecules that form the building blocks of proteins. OR Any of a class of 20 molecules that combine to form proteins in living things. OR The basic building block of a protein. There are about 20 different amino acids. Long chains of amino acids make up a protein.

Amplicon: An amplicon is a small piece of DNA that has been amplified by PCR. OR The product of PCR or LCR; a piece of DNA that has been synthesized using amplification techniques.

Amplification : An increase in the number of copies of a specific DNA fragment. OR Producing multiple copies of a chosen DNA region, usually by PCR (Polymerase Chain Reaction).

Analytical gel : The final gel in an RFLP test and the one that will be used to make the Southern blot.

Annealing : Annealing is a process where the forward and reverse primers anneal to separated DNA template strand. OR Annealing is a bimolecular process, but there are competing processes in which both intrastrand and interstrand base pairs can form, ultimately leading to the stable duplex

Antibiotic : Chemical that can be used to kill or inactivate bacteria within a person or animal. Antibiotics are widely used in medicine to treat diseases caused by bacteria. The first antibiotic discovered - penicillin - is produced naturally by some types of mould. Today there are many different types of antibiotic, many produced using the techniques of modern biotechnology.

Antibiotic resistance : The ability of bacteria to tolerate antibiotics and remain unaffected by them. Resistance may evolve naturally in bacteria after years of exposure to antibiotics. It is controlled by genes and can be spread between bacteria.

Antibody : A protein produced by the immune system which attaches to an antigen. When an antigen from outside is present in the body, it stimulates the production of a specific antibody that will combine with it, usually enabling it to be eliminated. There are many thousands of different types of antibodies.

Anticodon : Threenucleotide sequences complemen-tary to codons found on messenger RNA.

Antiparallel : Refers to two linear polymers that are opposite in orientation or polarity.

Arm: The stem-and-loop structure in a tRNA molecule that contains the sequence ribothymidylate–pseudouridylate–cytidylate

Asymmetric PCR : Asymmetric PCR is used to preferentially amplify one strand of the original DNA more than the other. It finds use in some types of sequencing and hybridization probing where having only one of the two complementary stands is ideal. PCR is carried out as usual, but with a great excess of the primers for the chosen strand. Due to the slow (arithmetic) amplification later in the reaction after the limiting primer has been used up, extra cycles of PCR are required.

Autorad : A film exposed by a radioactive tracer. In the case of forensic RFLP analysis the tracer is attached to a DNA probe which is in turn attached to a VNTR allele immobilized on a Southern blot.

Autosomal : Chromosomes which are not sex chromosomes.

Autosome : Any of the nonsexdetermining chromo-somes. Human cells have 22 pairs of autosomes. OR All chro mosomes except those involved in sex determi-nation. OR A chromosome not involved in sex determination. The human genome consists of 46 chromosomes, 22 pairs of autosomes (numbered 122), and one pair of sex chromosomes (the X and Y chromosomes).

B

Bacteria : Tiny onecelled organisms. OR Single celled organisms capable of reproduction and growth. Bacteria can be beneficial or harmful.

Bacteriophage : A virus which infects a bacterial cell. Also called phage.

Band : The exposed area on an autorad film that detects the presence of a particular DNA fragment. OR In gel electrophoresis, a band represents the dyed or stained molecule you're interested in. In a sample containing two different sized pieces of DNA, there will be two bands on the gel in that lane. The smaller piece of DNA will move more quickly through the gel, so it will be farther from the well. Each band is composed of hundreds of copies of the DNA molecule of that length.

Bandshift : The phenomena in which identical DNA fragments will move through an electrophoresis gel at different rates because of contamination or alterations to

Base (nucleotide) : The chemical units adenine (A), guanine (G), cytosine (C), and thymine (T) contained in DNA and that by virtue of their sequence determine the information content of the DNA. OR The chemical unit that characterises a nucleotide. In DNA, the bases found are adenine (A), guanine (G), thymine (T) and cytosine (C). In RNA, the bases are adenine, guanine, uracil (U) and cytosine. OR The molecular building blocks of DNA: adenine (A), cytosine (C), guanine (G), thymine (T). In DNA, A attaches only to T, and C attaches only to G. OR The basic building block of DNA. There are usually 4 bases - Adenine, Cytosine, Thymine and Guanine. A DNA strand incorporates thousands or millions of these 4 building blocks. The same building blocks are present in all forms of life.

Base pair (bp) : Two of the building blocks of DNA held together by weak bonds. In a DNA molecule, adenine always bonds with thymine (AT), and cytosine always bond s with guanine (CG). OR Two nitrogenous bases held together by weak bonds. In a DNA molecule adenine always pairs with thymine while cytosine always pairs with guanine. Thus, one DNA strand directs the synthesis of the other strand. This specificity of DNA base pairing is what enables accurate replication of a DNA molecule. OR Two nitrogenous bases paired together in double-stranded DNA by weak bonds; specific pairing of these bases (adenine with thymine and guanine with cytosine) facilitates accurate DNA replication; when quantified (e.g., 8 bp), refers to the physical length of a sequence of nucleotides

Base sequence: The order of bases in a DNA molecule. OR The order of nucleotide bases in a DNA molecule e.g AGTACGTA etc.

Base sequence analysis : A method, sometimes automated, for determining the sequence of bases in a strand of DNA.

Base substitution : One nucleotide (base) is replaced by another in a DNA molecule.

Baseline : Refers to a measure of the gene expression level of a gene or genes prior to a perturbation in an experiment, as in a negative control. Baseline expression may also refer to the expected or historical measure of expression for a gene.

Biomolecule : A minute particle of matter from a living or once living thing Chromosomes: Long chains of DNA that carry genes and other genetic Information. In eukaryotes they occur in pairs inside the nucleus of a cell.

Biotechnology : The use of biological processes to manufacture products. OR A broad term originally used to describe the application of biology in the creation of helpful products (for example, agriculture, brewing and baking were all considered types of biotechnology). Recently, the word has come to refer more to modern methods of using organisms and biological processes to create either genetically modified organisms or products (such as insulin and many pharmaceuticals) manufactured using the techniques of genetic engineering.

Blind testing : In a blind test, analysts do not know they are being tested. In most forensic DNA laboratories, blind tests are not used. In practice, it is almost impossible to design and implement an effective blind PT (proficiency test) program in forensic science. Most attempts have failed because they could not produce an effective case scenario with realistic representation of the prelaboratory steps. Others failed because the analyst recognized that the supposed "evidence" was a manufactured artifact. Overall, it has proven impossible to realize the theoretical extra benefits of blind testing, and resources have been devoted to promoting better quality external open tests.

Blot : See Southern blot.

Bt : Bacillus thuringiensis, a bacterium commonly found in soil. It produces a protein (Bt toxin) which is naturally toxic to some insects. Different Bt toxins (from different strains) affect different insect types.

Buffer (pcr buffer) : PCR buffer is necessary to create optimal conditions for activity of Taq DNA polymerase. Buffers often contain Tris-Hcl, KCl, and sometimes MgCl2. PCR buffers are often available in 10X concentration and are sometimes Taq formulation-specific.

C

Capillary electrophoresis (ce) : The platform for CE uses narrow silica capillaries (or tubes) containing a polymer solution through which the negatively charged DNA molecules migrate under the influence of a high voltage electric field. Important advantages of the CE technique, compared to slab gel electrophoresis, include quicker and more easily automated analyses

Capping : A process that messenger RNA undergoes aiding the molecule in binding and positioning itself on the ribosome.

Caps : (Cleaved Amplified Polymorphic Sequence) See PCR-RFLP

Carrier : A person who has a recessive mutated gene, together with its normal allele. Carriers do not usually develop disease but can pass the mutated gene on to their children.

Cdna(complementary DNA): DNA transcribed from an RNA molecule by an enzyme called reverse transcriptase. OR A DNA sequence which was produced from mRNA by reverse transcription. A cDNA is so-called because its sequence is the complement of the original mRNA sequence. However, when double-stranded cDNA is synthesized, it contains both the original sequence and its complement.

Cdna library : A collection of cDNA's, each of which has been inserted in a DNA vector (e.g. a circular DNA plasmid) and replicated in a bacterium such as E. coli. The bacteria maintain a ready pool of the cDNA's and can be cultured to make copies of the library for many experiments. A population of bacteria containing a single inserted cDNA is called a clone. To extract a cDNA from the library, the insert from the corresponding clone can be amplified using PCR primed off the vector sequences connected to the ends of the cDNA. OR A pool of complementary DNA clones produced by cDNA cloning of total messenger RNA from a single source (cell type, tissue, embryo).

Ceiling principle : An extremely conservative method of calculating the estimated occurrence of DNA profiles in forensic samples. It's purpose is to compensate for any undetected substructure that may exist in the populations that have been sampled.

Cell : The basic unit of structure and function in all living things.OR Small, watery, membranebound compartment filled with chemicals; the basic subunit of any living thing. OR The smallest component of life capable of independent reproduction and from which DNA is isolated for forensic analysis. OR The basic units of life. All living organisms are a collection of cells. Chromosomes The selfreplicating genetic structures of cells.

Cell division : The process by which cells reproduce. A cell divides by elongating and then pinching in the middle into two separate halves, each of which is a new cell. All multicellular organisms, which includes all plants and animals, started as a single cell. Repeated rounds of cell division resulted in the final organism.

Central dogma : The main principle of molecular biology, coined by Francis Crick, which states that genetic information flows from DNA to RNA to protein.

Chaintermination Reaction : When polypeptide synthesis is stop too early by an unwanted stop codon which resulted from a base mutation.

Charging enzymes : The enzymes that attach amino acids to the transfer RNA specific to them.

Chemical base: An essential building block. Dna contains four complementary bases adenine, which pairs with thymine, and cytosine, which pairs with guanine. In rna, thymine is replaced by uracil.

Chromosome : A structure made of DNA and protein. Most living things above the level of bacteria carry their DNA in the form of chromosomes. OR Structures, composed of DNA and protein, found in the nucleus of the cell. Most human cells contain 46 chromosomes;

23 derived from the mother and 23 derived from the father. Gametes (sperm or egg cells) contain 23 chromosomes until conception. OR A selfreplicating genetic structure, composed primarily of proteins and DNA, which carry nucleotide sequences known as genes. OR The units made up of proteins and DNA into which the DNA in cells is packaged. There are 23 pairs of chromosomes in each human cell.

Cistron : The section of messenger RNA that actually codes for the protein.

Clone : A group of identical genes, cells, or organisms derived from a single ancestor. OR A group of genetically identical cells or o rganisms that are descended from one parent. Identical twins are clones, as are colonies of bacteria that reproduce by simple cell division. OR A copy. Genes, cells or entire organisms can be cloned using the techniques of modern biotechnology. Usually, no two organisms, even if they belong to the same species, are genetically identical - but cloned organisms are. Some organism cloning occurs naturally, for example, when a new plant is formed from a cutting, or when humans produce identical twins.

Cloning : The process of making genetically identical copies. OR The process of asexually producing a group of cells (clones), all genetically identical, from a single ancestor. OR The procedure for purifying and reproducing in the laboratory specific small identical DNA sequences. OR Using specialized DNA technology to produce multiple, exact copies of a single gene or other segment of DNA to obtain enough material for further study. OR In molecular biology: the process of using DNA manipulation procedures to produce multiple copies of a single gene or DNA segment.

Cloning vector : DNA molecule originating from a virus, a plasmid, or the cell of a higher organism into which another DNA fragment of appropriate size can be integrated without loss of the vectors capacity for self- replication; vectors introduce foreign DNA into host cells, where it can be reproduced in large quantities. Examples are plasmids, cosmids, and yeast artificial chromosomes; vectors are often recombinant molecules containing DNA sequences from several sources.

Coding region : The portion of a DNA molecule that is transcribed into RNA.

Codis : Combined DNA Identification System, the FBI's database of DNA profiles of convicted offenders supplied to them by the states. OR Combined DNA Index System. A collection of databases of DNA profiles obtained from evidence samples from unsolved crimes and from known individuals convicted of particular crimes. Contributions to this database are made through State crime laboratories.

Codons : The basic unit of the genetic code, comprising threenucleotide sequences of messenger ribonucleic acid (mRNA), each of which is translated into one amino acid in protein synthesis.

Colinear : The term used to describe the linear relationship between the linear sequence of nucleotides in messenger RNA and the order of amino acids required to make the protein specified by that sequence of RNA nucleotides. For example, the sequence of amino acids in a protein corresponds to the sequence of nucleotide bases of the messenger RNA which was translated to make it.

Colony PCR : Colony PCR is can be used after a transformation to screen colonies for the desired plasmid. Primers are used which generate a PCR product of known size. Thus, any colonies which give rise to an amplification product of the expected size are likely to contain the correct DNA sequence. OR Use of a bacterial colony for direct polymerase chain reaction to amplify a specific nucleotide sequence.

Complementary base pair : A pair of bases in which the identity of one base defines the identity of its partner base. E.g. In a DNA molecule there are two complementary base pairsAdenine and thymine, and guanine and cytosine.

Complementary sequences : DNA sequences that form a doublestranded structure by matching base pairs; the complementary sequence to GTAC is CATG.OR DNA is made up of two long strands connected together along the middle (a bit like a zip), where bases on one strand are bonded to bases on the other side.If the base sequence is GTAC on one side, the other side is always CATG, thus complementing each other.

Constitutive gene : Or constitutive expression describes a gene that is transcribed continually compared to a facultative gene which is only transcribed as needed.

Contamination : The undesirable transfer of material to physical evidence (DNA) from another source.

Contig maps : Types of physical DNA maps that consist of overlapping segments of DNA (contigs) that, taken together, completely represent that section of the genome. (See Physical maps.)

Control samples : Cuttings, swabbings, etc., from unstained adjacent material. A control sample is material of a known source that presumably was uncontaminated during the commission of the crime (e.g., a sample to be used in laboratory testing to ensure that the surface on which the sample is deposited does not interfere with testing. For example, when a bloodstain is collected from a carpet, a segment of unstained carpet must be collected). The control sample should be taken adjacent to the biological stain being collected.

Conventional breeding : The techniques of animal or plant breeding that have been carried out for thousands of years. Conventional breeding usually involves choosing the individual plants or animals which possess the features closest to the desired ideal, and then breeding these individuals together. Conventional breeding is an approximate way of controlling gene combinations, and of ensuring that a gene or genes for a desirable trait are passed on or deliberately mixed with other desirable genes.

Cross contamination : The undesirable transfer of material between two or more sources of physical evidence.

Crossing over : A phenomenon, also known as recombi-nation, that sometimes occurs during the formation of sperm and egg cells (meiosis); a pair of chromosomes (one from the mother and the other from the father) break and trade segments with one another.

Cytoplasm : The cellular substance outside the nucleus in which the cell's organelles are suspended. OR The watery material that fills part of the cell; includes the organelles but not the nucleus. OR The viscid, semifluid matter contained within the plasma membrane of a cell, excluding the nucleus.

Cytosine : A nitrogenous base. Pairs with guanine in DNA molecules.OR A pyrimidine base; one of the four molecules containing nitrogen present in the nucleic acids DNA and RNA; designated by letter C. OR Designated by the letter C, it is a pyrimidine base. One of the four base molecules present in DNA.

D

DD-PCR (Differential display PCR) : Differential display PCR is a PCR-based technique that allows side-by-side comparison of multiple RNA samples and can facilitate the identification of both suppressed and induced genes. ORDifferential display PCR is a powerful technique that can be used to detect differences in gene expression between different types of cells and tissues.

Degenerate PCR : Degenerate PCR is in most respects identical to ordinary PCR, but with one major difference. Instead of using specific PCR primers with a given sequence, you use mixed PCR primers. That is, if you do not know exactly the sequence of the gene you are going to amplify, you insert "wobbles" in the PCR primers where there is more than one possibility. For instance, if you just have a protein motif, you can back translate the protein motif to the corresponding nucleotide motif. (Protein → Sequence).

Degradation : The fragmenting, or breakdown, of DNA by chemical or physical means. OR The breaking down of DNA by biological or chemical processes. In the context of forensic science these processes are usually the result of cell death or exposure to the elements.

Deletion : A mutation cause by the removal of one or more nucleotides from a gene or chromosome. OR A genetic mutation involving the loss of DNA. This may be small, affecting only a portion of a single gene, or it may be large, such as partial or complete loss of a chromosome, affecting many genes.

Denaturation : Use of heat or alkali to disrupt hydrogen bonds that hold the two strands of the DNA double helix together. OR With respect to nucleic acids, refers to the conversion from double-stranded to the single-stranded state, often achieved by heating or alkaline conditions. This is also called "melting" DNA. OR The process of splitting the complementary double strands of DNA to form single strands. OR Describes the conversion of DNA from the double-stranded to the single-stranded state; separation of the strands is most often accomplished by heating.

Deoxynucleotide triphosphates (dntps) : dNTPs are the nucleotide bases added to the growing DNA strand by the DNA polymerase. The concentration of each dNTP in the reaction mixture is usually 200μM. It is very important to have equal concentrations of each dNTP (dATP, dCTP, dGTP, dTTP), as inaccuracy in the concentration of even a single dNTP dramatically. OR increases the misincor-poration level. Or the bases added to a primer during the PCR that comprise the newly synthesized strand. Examples of dNTPs are dATP, dUTP, dCTP, dGTP and dTTP.

Dideoxy sequencing : Dideoxynucleotide sequencing, also known as the "Sanger method," is a technique which uses dideoxyribose instead of deoxyribose to stop the synthesis of a complementary DNA strand at various points when sequencing.

Differential extraction : A procedure in which sperm cells are separated, or extracted, from all other cells in a sample.

Differentiation : The process of change from an unspecialised cell into a specialised one. The cell's structure and function may change. In vertebrates, most cells in the embryo are not yet differentiated. During embryonic development, cells differentiate into different types, such as nerve, muscle or blood cells. Once differentiated, these cells cannot usually change back into 'generalised' cells.

Diploid : Two complete genomes in every cell, one derived from mom and the other from dad. OR Having a full set of genetic material consisting of paired chromosomes that contain one chromosome from each parent.

Dna : The substance of heredity; a large molecule that carries the genetic information that cells need to replicate and to produce proteins. OR Deoxyribonucleic acid. Dna is the double helix molecule of hereditary. It is composed of 4 bases or building adenine (a), thymine (t), cytosine

(c) and guanine (g). These bases are strung along two opposing strands, like pearls on a string. The a is always bound to t and g is always bound to c by week hydrogen bonds. The specific order of bases is the dna sequence. OR Deoxyribonucleic acid, the biological polymer that stores the genetic information in all free living organisms. Two linear molecules entwine to form the double helix.

DNA amplification : DNA amplification is a process for replicating large amounts of DNA from just a few original strands. Thermal Gradient's technology greatly simplifies and accelerates the polymerase chain reaction (PCR), the most popular DNA amplification method available. OR The creation of numerous copies of a given DNA sequence, generally through the polymerase chain reaction (PCR). OR DNA amplification is a process for making large amounts of DNA from just a few original strands.

Dna analysis : The process of testing to identify DNA patterns or types. In the forensic setting, this testing is used to exclude or include individuals as possible sources of body fluid stains (blood, saliva, semen) and other biological evidence (bones, teeth, hair). This testing can also be used to indicate parentage.

DNA fingerprinting : A process which uses fragments of DNA to identify the unique genetic makeup of an individual. OR A technique for identifying individual organisms based upon the uniqueness of their DNA pattern. The technique has applications in forensics, paternity testing, anthropology, conservation biology and ecological research. OR The use of DNA sequences specific for an organism for identification purposes.

DNA mixtures : A sample that contains the DNA of more than one individual.

Dna polymerase : An enzyme that aids in DNA replication. It is the major enzyme in the process. OR An enzyme that produces or synthesizes DNA. These enzymes always use an existing DNA molecule as a template for producing a new strand of DNA. OR An enzyme that produces or synthesizes DNA. These enzymes always use an existing DNA molecule as a template for producing a new strand of DNA.

DNA probe : A short piece of DNA made in the laboratory by chemical or biological means that can be labeled with a tracer and used to identify a specific gene. A specific gene is identified by a specific DNA probe. OR To find the location of a gene along a stretch of DNA, scientists may construct a DNA probe. This consists of a strand of DNA that is complementary to a portion of the DNA in the gene being sought.

Dna profile : The result of determining the relative positions of DNA sequences at several locations on the molecule. Each person (except identical twins) has a unique DNA profile when used in the context of the CODIS database, which evaluates 13 specific DNA locations.

DNA profiling or typing : A variety of DNA tests used by forensic scientists to aid in the identification of the individual source of evidence or to resolve a parentage question.

DNA repair genes : Certain genes that are part of a DNA repair pathway; when altered, they permit mutations to pile up throughout the DNA.

DNA replication : The use of existing DNA as a template for the synthesis of new DNA strands.

DNA sequencing : The process of determining the order of bases in a segment of DNA.OR Determining the exact order of the base pairs in a segment of DNA. OR The relative order of base pairs in any sample of DNA. See base sequence analysis.

DNA sequencing : Determining the order of bases in a segment of DNA.

Domi nant gene : A gene which, when present on a chromosome, passes on a certain physical characteristic, thereby dominating over a recessive gene present on another chromosome. Double helix A common name for DNA; it refers to the molecule's doublestrande d, spiraling structure.

Dominant allele : A gene that is expressed, regardless of whether its counterpart allele on the other chromosome is dominant or recessive. Autosomal dominant disorders are produced by a single mutated dominant allele, even though its corresponding allele is normal. (see recessive allele.)

Double helix : The shape the DNA assumes after it replicates during cell life. OR The familiar shape that two linear strands of DNA with complementary sequences assume when bonded together. OR The shape that two linear strands of DNA assume when hydrogen-bonded together.

Double-stranded (ds) : Consisting of two bound strands, each of which is the complement of the other. DNA is usually double-stranded, while mRNA is not.

Down regulation : Is decreasing the rate of gene expression.

Downstream : In the direction of the 3' end of a DNA strand. OR Toward the 3' end of a nucleotide sequence.

Dsdna : (double stranded DNA)

Dys : D = DNA Y = Chromosome; S = Single copy sequence. The DYS numbering scheme (e.g. DYS388, DYS390) for the YSTR haplotype markers are controlled and administered by an international standards body called HUGO Human Gene Nomenclature Committee based at University College, London. OR DNA + Y chromosome +

[unique DNA] Segment. DYS numbers are assigned by HUGO (Human Genome Organization).

E

Ecoli : A species of bacterium that lives naturally in human and other mammal intestines, and which is widely used in studies of molecular biology and for research into genetic engineering.

Electropherogram : The graphic representation of the separation of molecules by electrophoresis or other means of separation. OR The pattern of fluorescent peaks produced by alleles during electrophoresis. Collectively, this pattern is the dna profile.

Electrophoresis : A method of separating large moleculessuch as DNA fragments or proteinsfrom a mixture of similar molecules. OR The process of separating charged molecules, such as fragments of DNA, in an electric field. The electric current is passed through an agarose gel, which allows the DNA fragments to separate, by size. OR A method of separating large molecules (such as DNA fragments) from a mixture of similar molecules. An electric current is passed through a medium at a different rate, depending on its electrical charge and size. Separation of DNA markers is based on these differences. OR The process of moving DNA or proteins through a sieving matrix that will separate them by size. This process is possible because most biological molecules have an electric charge and so will move in response to the electrical field that is applied to the separation matrix. OR A method of separating large molecules (such as DNA fragments or proteins) from a mixture of similar molecules. An electric current is passed through a medium containing the mixture, and each kind of molecule travels through the medium at a different rate, depending on its electrical charge and size. Separation is based on these differences. Agarose and acrylamide gels are the media commonly used for electrophoresis of proteins and nucleic acids. OR A technique that separates DNA fragments on the basis or their size by running them through a viscous material. The fragments move through the material due to the application of an electric current which attracts the negatively charged DNA to the positive terminal, and the fragments are separated due to the fact that the smaller fragments will move through the matrix faster than the larger fragments and will thus arrive sooner at the positive terminal.

Elimination/Reference samples : A term used to describe a sample of known source taken for comparison purposes.

Elongation : The second step of translation in which he protein being manufactured begins to grow in size.

Elongation or extension : Phase of PCR cycle following annealing of primer during which the Taq polymerase manufactures a strand of DNA. The optimum temperature depends on the enzyme used but is usually between 68-72° C. OR Refers to the elongation of the DNA chain that is being synthesized using the parent DNA strand as the template for synthesis of that daughter strand. This is a natural process that occurs during DNA replication. Extension occurs during the PCR process with DNA polymerases.

Embryo : The stage of an organism's development from the first division of the zygote (to give two cells) until the development of organs.

Enzymes : Proteins that serve as catalysts for specific biological reactions. A catalyst is a substance that helps a chemical reaction occur. It remains unchanged itself and so is required in trace amounts. The two most commonly used in DNA testing are the restriction endonuclease Hae III and Taq DNA polymerase.

Ethidium Bromide (etbr) : Intercalates within the structure of nucleic acids in such a way that they fluoresce under UV light. Ethidium bromide staining is commonly used to visualize RNA or DNA in agarose gels placed on UV light boxes. Proper precautions are required, because the ethidium bromide is highly mutagenic and the UV light damaging to the eyes. OR A dye that stains DNA and is illuminated by ultraviolet or black light on a device called a transilluminator. It is used to visualize DNA in gels following electrophoresis and can be added to the DNA before or after.

Evolution : Process by which a species develops into new and different groups in response to environmental selection favouring the reproductive success of certain individuals The fossil record reveals evolutionary history over long periods of time and shows the modification of species as they have adapted to suit different environments.

Exclusion : When the DNA from a crimes scene fails to match that of a suspect. Inclusions are probability statements, exclusions are absolute. OR A DNA test result indicating that an individual is excluded as the source of the DNA evidence. In a criminal case, "exclusion" does not necessarily equate to "innocence." This occurs when one or more types from a specific location in the DNA of a known individual are not present in the type(s) for that specific location in the DNA obtained from an evidence sample. OR In forensic and paternity testing the situation that exists when testing has demonstrated that a particular sample could not have come from an individual who was also tested or that the alleged father in a paternity case cannot be the father of the child.

Exogenous DNA : DNA originating outside an organism that has been introduced into the organism.

Exon : Areas in messenger RNA that code for proteins. OR Part of a gene whose sequence is present in a mature mRNA after splicing.

Expressed sequence tag (est) : A small sequence from an expressed gene that can be amplified by PCR. ESTs act as physical markers for cloning and full length sequencing of the cDNAs of expressed genes. Typically identified by purifying mRNAs, converting to cDNAs, and then sequencing a portion of the cDNAs. OR A partial sequence of a randomly chosen cDNA, obtained from the results of a single DNA sequencing reaction. ESTs are used to identify transcribed regions in genomic sequence, to characterize patterns of gene expression in the tissue that was the source of the cDNA and as markers for genetic mapping.

F

Facultative gene : Is a gene which is only transcribed as needed compared to a constitutive gene.

FISH (fluorescence in situ hybridization) : A mapping technique that uses fluorescent tags to identify specific locations of chromosomes.

Flanking regions : The DNA sequences extending on either side of a specific gene or locus. OR For microsatellites, the flanking regions are the stretches of DNA outside the simple sequence tandem repeat (STR).

Fluorescence : Emission of light by excited molecules as they revert to the ground state. OR Emission of secondary light generated by excitation by ultraviolet light, a beta particle or a gamma ray. OR Molecules that are fluorescent emit light upon excitation with light. In practice, dna molecules are tagged with fluorescent molecules. Upon excitation with laser light, fluorescent scanners (such as the fmbio ii or abi 377) or capillary electrophoresis instruments (abi 310) can detect and size the tagged dna. Since it is possible to tag dna with different fluorescent molecules that emit light in different regions of the spectrum, it is possible for instruments to detect and distinguish between dna fragments of similar size.

Forensic science : The application of science to analyze evidence involved in criminal and civil litigation. OR The application of scientific knowledge to questions of civil and criminal law, especially in court proceedings.

Formamide : Amplified DNA is mixed with formamide to retain DNA in single stranded state. Poor quality formamide have been associated with DNA degradation and artifacts on the ABI 310. These effects are not generally seen on polyacrylamide gel based systems.

Fractions : The result of the differential extraction; separating sperm cells from all other DNA material.

Functional gene tests : Biochemical assays for a specific protein, which indicates that a specific gene is not merely present but active.

G

Gel : A dense net work of fine particles dispersed with water (jello is a gel). Used to separate differentsized strands of DNA. OR Polyacrylamide or agarose used in DNA electrophoresis; provides a sieving action on DNA fragments as they migrate in an electric field. OR A semisolid matrix used in the electrophoretic separation of DNA and proteins. The two most commonly used in DNA testing are agarose and polyacrylamide.

Gene : The fundamental unit of heredity. A unit of inheritance; a working subunit of DNA. Each of the body's 20,000 to 25,000 genes contains the code for a specific product, typically, a protein such as an enzyme. (Revised October 2004). OR Unit of heredity, region of DNA that encodes or regulates the expression of a protein. OR The basic unit of heredity; a sequence of DNA nucleotides on a chromosome. OR

Gene deletion: The total loss or absence of a gene.

Gene expression : The process by which a gene's coded information is translated into the structures present and operating in the cell (either proteins or rnas). OR Is the transcription of DNA into messenger RNA by RNA polymerase. The messenger RNA is then translated into protein by the ribosome. OR The process by which genes express themselves in the cell, ge ne expression results in the manufacture of proteins that determine an organism's characteristics.

Gene families : Groups of closely related genes that make similar products.

Gene mapping : Locating the relative position and the distances between genes within a chromosome. OR Determining the relative positions of genes on a chromosome and the distance between them. OR Determining the relative positions of genes on a DNA molecule.

Gene markers : Landmarks for a target gene, either detectable traits that are inherited along with the gene, or distinctive segments of DNA.

Gene pool : The total number of different genes in a population of organisms of the same species.

Gene product : The biochemical material, either RNA or protein, resulting from expression of a gene.

Gene silencing : Genes can be switched off by activating a gene switch that codes for a repressor. The repressor molecule will prevent the activity of the target gene. Genes can also be silenced if the DNA of which they are made is subjected to special chemical changes that may prevent it from being read. A new process called RNA interference is also now known to silence or block individual genes.

Gene technology : The ability to manipulate, modify and transfer genes or segments of DNA.

Gene testing : Examining a sample of blood or other body fluid or tissue for biochemical, chromo-somal, or genetic markers that indicate the presence or absence of genetic disease.

Gene therapy : Treating disease by replacing, manipulating, or supplementing nonfunctional genes. OR A newly evolving technique used to treat inherited genetic diseases. The medical procedure involves substituting the defective gene in the cells of a patient's body with a healthy gene.

Genetic code : The code in which the instructions of life are written. The genetic code refers to the sequence of bases in a DNA molecule. There are four possible bases, and their sequence spells out how to build proteins. In turn, the proteins are responsible for constructing and operating the features of the organism.

Genetic diversity : The variety of different genes occurring within a population or species. In most species, many different versions or "alleles" of a gene will exist in a population of different individuals. For example, in humans, everybody has a gene affecting eye colour, but the different versions of these genes give rise to variety in eye colour in the population.

Genetic engineering : Another word for gene technology. OR The technology used to genetically manipulate living cells to produce new chemicals or perform new functions.

Genetic genealogy : The new field of family history research utilising the Ychromosome and mtdna to prove or disprove biological connections. The Ychromosome links the strictly paternal line (father to son), whilst mtdna links the mother to offspring line (mother to sons and daughters). For the vast majority of family history studies, the Ychromosome is studied.

Genetic loci : Places in the genetic material of an organism where specific DNA sequences can be found.

Genetic modification : The deliberate changing of the genetic material in an organism. Scientists can determine whether or not the change will be passed onto offspring. Usually in gmos, the modification is passed on. Genetic modification is a general term that can cover many processes. (It is possible to modify genes and not have the modification passed on to offspring.)

Genetic switches : Genes can be 'switched' on or off (activated or inactivated). Certain regions of DNA control the activation of genes.

These regions – or the protein molecules they produce are sometimes referred to as genetic switches.

Genetic testing (or genetic screening) : A technique used to determine whether a person or organism has a certain gene.

Genetically modified organism (GMO) : An organism with genetic material that has been altered by genetic engineering (or gene technology).

Genetics : The study of the patterns of inheritance of specific traits. OR The study of heredity and variations in living organisms. The term 'molecular' genetics is used to describe the study of genes and their function (i.e. Genetics at a molecular level).

Genome : All the genetic material in the chromosomes of a particular organism. OR The total genetic material of an individual or species. OR All of the genetic material or DNA in a cell. OR All the genetic material in the chromosomes of a particular organism. The human genome consists of three billion bases, organized in about 100,000 genes on 23 chromosomes.

Genome maps : Charts that indicate the ordered arrangement of the genes or other DNA markers within the chromosomes.

Genome projects : Research and technology development efforts aimed at mapping and sequencing some or all of the genome of an organism. The improvements and lower cost of technology that genome projects have brought make possible the new field of genetic genealogy for genealogists.

Genotype : The genetic constitution of an organism. OR The actual genes carried by an individual (as distinct from phenotypeñthat is, the physical characteristics into which genes are translated). OR The entire genetic identity of an individual, including alleles, or gene forms, that do not show as outward c haracteristics.

Germ cells : The reproductive cells of the body, either egg or sperm cells.

Guanine (G) : A purine base; one of the four molecules containing nitrogen present in the nucleic acids DNA and RNA; designated by letter G. OR Designated by the letter G, it is a purine base. One of the four base molecules present in DNA. OR A nitrogenous base. Pairs with cytosine in DNA molecules. OR An organic base, one of the four monomers or building blocks from which DNA is made.

H

Haploid : A single set of chromosomes (half the full set of genetic material), present in the egg and sperm cells o f animals and in the egg and pollen cells of plants. Human beings have 23 chromosomes in their sex cells. Compare to diploid.

Haplotype diversity : A measure of the uniqueness of a particular haplotype in a given population.

Helicase : An enzyme that separates the strands of a DNA molecule prior to replication.

Herbicide : A chemical effective at killing plants. Widely used in agriculture, horticulture and gardening to control unwanted plants (referred to as weeds). Herbicides can be selective, killing only some plant species or a group of plants, or nonselective (sometimes called broad spectrum) – which are generally poisonous to all plants.

Hereditary mutation : A gene change in the body's reproductive cells (egg or sperm) that becomes incorporated in the DNA of every cell in the body; also called germline mutation. (See Acquired mutations.)

Heredity : The transmission of characteristics from one generation to the next. OR The transmission of characteristics (hair colour, facial features, genetic disorders etc.) From one generation to the next. OR The phenomena whereby biological traits are passed from one generation to the next. Heredity results from parents passing genes to their offspring and is why related family members tend to have similar physical characteristics.

Heteroduplex : DNA A double-stranded DNA molecule or a DNA/RNA hybrid where each strand is from a different source. OR A DNA double helix formed by annealing single strands from different sources; if there is a sequence difference between the strands, the heteroduplex may show single strand loops or bubbles (unpaired regions).

Heteroduplex analysis : The study of the mobility of heteroduplex DNA under polyacrylamide gel electro-phoresis. The reduced mobility of heteroduplex DNA compared with homoduplex DNA is proportional to the degree of divergence of the sequences.

Heteroplasmy : The presence of more than one mtDNA type within a single individual.

Heterozygosity : In forensic science often used to describe the percentage of individuals in a population database who are heterozygotes.

Heterozygote : An individual who has inherited and so carries different alleles of a gene.

Hla : Human leukocyte antigen, a substance found on the surface of most cells that differs from person to person. One of these, the DQA1 gene, is examined by the most widely used PCR based forensic DNA test.

HLA dqalpha : A polymorphic gene in the Human Leukocyte Antigen (HLA) region of chromosome 6 that has been well studied and analyzed for many purposes including paternity testing, transplantation biology, and human DNA identification testing.

Homoduplex : DNA A double-stranded DNA molecule where both strands are from the same source.

Homologies : Similarities in DNA or protein sequences between individuals of the same species or among different specie s.

Homozygote : An individual who has identical alleles of a gene. An excess of homozygosity in a population database, according to the HardyWeinberg Principle, would indicate that there are problems with the data and that it should not be used.

Hot-start PCR : Is a modification of conventional PCR that reduces non-specific product amplification. In this procedure amplification cannot occur until the reaction temperature is above that where non-specific annealing of primers to targets occurs. This block in amplification is usually accomplished by using a DNA polymerase that is inactive until higher temperatures are reached.

Housekeeping gene : Is typically a constitutive gene that is transcribed at a relatively constant level across many or all known conditions. The housekeeping gene's products are typically needed for maintenance of the cell.

Human genome : The full collection of genes needed to produce a human being.

Human genome project (hgp) : An international research effort (led in the United States by the National Institutes of Health and the Department of Energy) aimed at identifying and ordering every base in the human genome. OR A worldwide project aimed at deciphering all the three billion bases of the human genome, including mapping and sequencing every gene. This information will help to more rapidly identify genes causing diseases in humans.

Hybrid: Offspring resulting from the cross of two different varieties or species. The greater the genetic distance between the parents (that is, the more different the parents are), the more likely the hybrid is to be sterile. A mule is an example of a hybrid, and results from the cross of a donkey and horse. Hybrids, especially fertile hybrids, occur much more readily in plants than in animals. An example of a commonly used hybrid plant is wheat, which contains the genes of three closelyrelated plants.

Hybridization: The joining together of complementary strands of DNA by the process of base pairing when the two strands are from different sources. For example, a DNA strand immobilized on a blot and its complementary radioactively labeled DNA probe. OR The process of joining two complementary strands of DNA or one each of DNA and RNA to form a double-stranded molecule. One strand is often labeled and used as a probe to detect the presence of the second strand. OR The joining together of complementary strands of DNA by the process of base pairing when the two strands are from different sources. For example, a DNA strand immobilized on a blot and its complementary radioactively labeled DNA probe. OR In molecular biology: the binding of sequences to form a double-stranded structure.

Hydrogen bonding : Intermolecular forces that hold the nitrogenous bases of nucleotides on both strands of the DNA molecule forming the "steps" of the spiral staircase known as the DNA double helix.

Hypervariable : An area on the DNA which can have many different alleles in differing sequences. OR A DNA locus that shows extremely high variation in people. Many alleles exist in the population.

I

Imprinting : A biochemical phenomenon that determines, for certain genes, which one of the pair of alleles, the mother's or the father's, will be active in that individual.

***In situ* PCR :** *In situ* PCR is a method in which the polymerase chain reaction actually takes place in the cell on a slide, and the product can be visualized in the same way as in traditional in situ hybridization. OR In situ PCR is a powerful technique in which gene fragments are amplified within tissue sections.

***In vitro*:** Outside a living organism. OR Inside a living organism.

Inborn errors of metabolism : Inherited diseases resulting from alterations in genes that code for enzymes.

Inclusion : A probability statement that describes a DNA match between a crime specimen and suspect. OR In forensic and paternity testing the situation that exists when testing has demonstrated that a particular individual belongs to the group of individuals who could be the source of the sample or in a paternity case the father of the child. They have not been excluded.

Inclusion (Failure to Exclude) : The inability to exclude an individual as a possible source of a biological sample. This occurs when all types from a specific location in the DNA of a known individual are also present in the types for that specific location in the DNA obtained from an evidence sample.

Iniation codon : the codon (AUG) that initiates protein synthesis.

Initiation : The first step of translation in which the process of protein synthesis begins.

Insertion : A type of mutation in which one or more nucleotides is inserted into a DNA sequence. Small insertions within a gene can alter the reading frame, and thus the amino acid sequence of the encoded protein. OR Addition of a DNA element into a chromosome.

Intercalating dyes (SYBR Green and similar) : Real Time Amplification System. Despite being the most economical option for detection and quantifying real time amplified products, intercalating dyes (i.e. SYBR Green), bind to any double-stranded DNA present in the reaction, including primer-dimers and other nonspecific reaction products. This requires to take specific precautions. Therefore most of the protocols for kits based upon this technology include a melting curve analysis, intended to identify the amplified product that has been detected.

Intron : Part of a gene whose sequence is transcribed but not present in a mature mRNA after splicing. OR DNA sequence that interrupts the protein-coding sequence of a gene; an intron is transcribed into RNA but is cut out of the message before it is translated into protein. OR The DNA base sequence interrupting the protein-coding sequences of a gene; intron sequences are transcribed into RNA but are cut out of the message before it is translated into protein. OR Areas in the middle of the messenger RNA molecule that do not code for proteins; these sequences are transcribed into RNA but are cut out of the message before it is translated into protein.

Inversion : A type of mutation results in an inverted segment in the chromosome. OR A type of mutation in which a length of DNA is broken in two positions and repaired in such a way that the medial segment is now present in reverse order. Inversions range in size from those large enough to be visible cytogenetically to those involving only a few base pairs. OR The term introduced by Sturtevant (1926) for the reversal of the linear sequence of the genes in a segment of a chromosome. In an inversion heterozygote one of the chromosomes contributed by one parent has an inverted segment while the homologous chromosome contributed by the other parent has the normal gene sequence.

Ipcr (inverse pcr) : Variation of PCR that makes the amplification of DNA segments of unknown sequence that flank DNA segments of known sequence possible; in brief, total DNA is digested to completion and fragments ligated under conditions that favour circularization of fragments; pair of PCR primers, designed from known sequence, are used to prime PCR from opposite strands resulting in amplification of fragment of unknown sequence. OR Inverse PCR is a method used to allow PCR when only one internal sequence is known. This is especially useful in identifying flanking sequences to various genomic inserts. This involves a series of digestions and self ligation before cutting by an endonuclease, resulting in known sequences at either end of the unknown sequence.

Irresolvable mixture : A DNA profile where multiple individuals have contributed biological material and no profile is more or less apparent than any other and the developed alleles cannot be isolated to a single source.

Isotope: The radioactive form of an atom known as a radioisotope.

J

John doe warrant : Stretches of DNA that do not code for genes; "most of the genome consists of junk DNA." OR A warrant used when crime scene evidence yields a DNA profile but the individual corresponding to the DNA profile is unknown. In lieu of the suspect's name, the warrant will be filed as "John Doe" and cite the DNA profile.

Junk DNA : Most of the DNA in our bodies is termed 'junk DNA'. It is noncoding (isn't used for making proteins) and thus does not control medical function but is probably involved in genomic organization. Genetic genealogy only looks at the markers within junk DNA and thus the YDNA test cannot derive any medical information. OR In molecular biology, "junk" DNA is a collective label for the portions of the DNA sequence of a chromosome or a genome for which no function has yet been identified. About 98.5% of the human genome has been designated as "junk", including most sequences within introns and most intergenic DNA. While much of this sequence is probably an evolutionary artifact that serves no present-day purpose, some may function in ways that are not currently understood.

K

Karyotype : A photomicrograph of an individual's chromosomes arranged in a standard format showing the number, size, and shape of each chromo some type.

Kilobase (kb) : Unit of length for DNA fragments equal to 1000 nucleotides. OR Unit of DNA or RNA sequence equal to 1000 nucleotides.

L

Ladder: A mixture of the most common STR alleles that are used to compare with the amplified samples.

Lane : In gel electrophoresis, a lane is the section of a gel below one well. The gels we are using have six wells, so they have six corresponding lanes. When you look at a stained gel, all the bands in one lane came from the same sample of DNA loaded into one well.

Ldis : The Local DNA Index System of CODIS, which uploads forensic and offender DNA profiles to the State DNA Index System, or SDIS.

Leader : The section of messenger RNA before the iniation codon that is not translated during protein synthesis.

Leading daughter strand : The DNA strand that is synthesized continuously during replication Compare lagging strand.

Length heteroplasmy : The presence of mtDNA molecules that differ in length.

Library: In molecular biology, a "library" is a complex mixture of recombinant DNA molecules in a suitable cloning vector representing either the entire genome of an organism (a genomic library) or the messenger RNA population of a whole organism, cell type, or tissue type (a cDNA library). OR An unordered collection of clones (i.e., cloned DNA from a particular organism) whose relationship to each other can be established by physical mapping.

Ligase : An enzyme responsible uniting the Okazaki fragments of the lagging strand during DNA replication.

Ligase chain reaction (LCR) : A method of DNA amplification similar to PCR. LCR differs from PCR because it amplifies the probe molecule rather than producing amplicon through polymerization of nucleotides. Two probes are used per each DNA strand and are ligated together to form a single probe. LCR uses both a DNA polymerase enzyme and a DNA ligase enzyme to drive the reaction. Like PCR, LCR requires a thermal cycler to drive the rxn and each cycle results in a doubling of the target nucleic acid molecule. LCR can have greater specificity than PCR.

Linkage : When two or more alleles are inherited together, often because of their location on the chromosome. OR A measure of association between two loci. Loci on different chromosomes are nonlinked. Those that are close together on the same chromosome are closely linked and are likely to be inherited together.

Linkage analysis : A genehunting technique that traces patterns of heredity in large, highrisk families, in an attempt to locate a diseasecausing gene mutation by identifying traits that are coinherited with it.

Linkage disequilibrium : A specific allele of one locus being found with a specific allele of another locus more often than would be expected by chance. Occurs when the two loci are close together on the same chromosome. They are then said to be linked because they will tend to be inherited together.

Linkage equilibrium : When two or more genetic loci appear to segregate randomly in a given population. The genotypes appear randomly with respect to each other.

Locus : A specific location on a chromosome (loci, plural). OR The specific physical location of a gene on a chromosome. OR The position of a gene, DNA marker or genetic marker on a chromosome. OR Latin for place. See marker. OR A specific physical position on a chromosome at which a gene or gene pair resides.

Long PCR : PCR in which the product is several kilobases long. Using cocktails of enzymes products of up to 40 kb have been accurately manufactured. OR Long PCR (i.e. amplification of DNA fragments of several thousand base pairs) is now widely used in numerous fundamental or applied scientific fields and represents an invaluable diagnostic tool in molecular medicine.

Low copy number : Refers to examination of less than 100pg (picograms) of input DNA.

M

Mage : Stands for microarray and Gene Expression, a group that "aims to provide a standard for the representation of microarray expression data that would facilitate the exchange of microarray information between different data systems".

Magnesium : Magnesium is a required cofactor for thermostable DNA polymerases, and magnesium concentration is a crucial factor that can affect the success of the amplification. OR DNA polymerase requires magnesium for activity. Magnesium is usually supplied to a PCR amplification in the form of magnesium chloride.

Major contributor profile : A DNA profile where multiple individuals have contributed biologic material and one individual's DNA profile is more apparent.

Mapping : See gene mapping, physical map.

Marker: An identifiable physical location on a chromosome (e.g., restriction enzyme cutting site, gene) whose inheritance can be monitored. OR Also called a Locus (pl. Loci). A gene of a known location on a chromosome. See STR marker. The DNA Heritage YDNA test uses 21 STR markers. OR Pieces of DNA sequence of known locations on chromosomes that are used to identify the specific genetic variations an individual possesses. OR

1. A genetic marker is a gene that has been mapped and can be identified.
2. A molecular weight marker is a DNA fragment of known size used as a comparison standard in estimating the size of a DNA fragment of unknown size. ORAn identifiable physical location on a chromosome (e.g., restriction enzyme cutting site, gene, minisatellite, microsatellite) whose inheritance can be monitored. Markers can be expressed regions of DNA (genes) or some segment of DNA with no known coding function but whose pattern of inheritance can be determined. OR A gene of known location on a chromosome and phenotype that is used as a point of reference in the mapping of other loci.

Marker gene : A gene, whose presence is easily detectable, which is inserted into a GMO along with the desired gene. The presence of the marker gene allows scientists to know that the insertion of the desired genes has been successful.

Match : Genetic profiles are said to "match" when they have the same allele designations at every loci.

Megabase (mb) : Unit of DNA or RNA sequence equal to one million nucleotides.

Meiosis : The production of se x cells, which are not genetically identical, through a series of cell divisions. Compare to mitosis. OR The process of cell division in which a single cell produces four daughter cells each of which contains half of the number of chromosomes of the parent cell.

Melanoma : A cancer that begins in skin cells called melanocytes and spreads to internal organs.

Melting of DNA : Denaturation of duplex DNA by heat or increased pH leading to strand separation.

Membrane : The solid support, usually nylon membrane, to which the DNA is transferred and affixed during the Southern blotting process.

Messenger RNA (mrna) : A singlestranded molecule of ribonucleic acid that directs protein production. OR An RNA molecule transcribed from the DNA of a gene, and from which a protein is translated by the action of ribosomes. The basic function of the nucleotide sequence of mRNA is to determine the amino acid sequence in proteins. OR RNA that serves as a template for protein synthesis or for synthesis of cDNA.

Mged : Is the Microarray Gene Expression Data Society, which "aims to facilitate the sharing of microarray data generated by functional genomics and proteomics experiments".

Microb organism : A microscopic living thing, such as all bacteria and viruses, many types of fungi and other singlecelled life forms.

Microbial epidemiology : Microbial relating to a microbe or microbes; epidemiology the study or the distribution and determinants of healthrelated states or events in specified populations, and the application of this study to control health problems.

Microgram (μg) : One millionth of a gram. OR A measurement of weight. One microgram is equal to one thousandth of a milligram.

Microsatellite : Microsatellite Marker. A short (up to several hundred base pairs) segment of DNA that consists of multiple tandem repeats of a two or three base-pair sequence. Microsatellites expand and contract (that is, add or remove repeat units) with a frequency much higher than other types of mutations, making them useful as polymorphic markers in closely related mouse strains. OR

Microsatellites are defined as short patterned non-coding regions containing specific genetic sequences that are repeated one right after the other (in direct sequence) within the genome of an organism. The repeated sequence in a microsatellite consists of two, three or four nucleotides (di-, tri-, or tetra-nucleotide repeats respectively) and can be repeated many times (upwards of 100). OR Repetitive stretches of short sequences of DNA used as genetic markers to track inheritance in families. They are short sequences of nucleotides (example: ATCG) which are repeated over and over again a number of times in tandem. Changes sometimes do occur, however, and the number of repeats may increase or decrease. OR A type of repetitive DNA based on very short repeats such as dinucleotides, trinucleotides or tetranucleotides.

Mitochondria : Organelles that exist inside cells and that contain a small amount of DNA (16,000 bp). A small region of the mitochondrial DNA varies from person to person. Because a cell may contain many thousands of mitochondria there are correspondingly many more copies of the mitochondrial DNA available for testing. molecular weight or size marker. See marker. OR An organelle (a specialised structure) in the cytoplasm of nearly all eukaryotic cells containing enzymes important for cell metabolism including those responsible for the conversion of food to usable energy Mitochondria are thought to once have been bacteria that were incorporated in the ancestral eukaryote cell and they contain the remnants of a simple bacterial genome.

Mitochondrial DNA (mtdna) : He DNA found in the many mitochondria found in each cell of a body. The sequencing of mitochondrial DNA can link individuals descended from a common female ancestor. OR The genetic material found in mitochondria, the organelles that generate energy for the cell. Not inherited in the same fashion as nucleic DNA. OR **mtDNA** The circular DNA contained inside the mitochondria. The mitochondria are small organelles residing in animal cells which provide the power to the cell. They occupy about onefifth of each cell. The mtdna is passed from mother to her offspring (both sons and daughters), but only the daughter will pass it on. The DNA sequence can be read and compared against a standard sequence (the Cambridge Reference Sequence) and deep (i.e. Several thousands of years), but very broad genealogies can be deduced. A mtdna haplogroup can usually be assigned for any given sequence.

Mitosis : The production of cells that are genetically identical to the original cell. Compare to meiosis. OR The process of nuclear division in cells that produces daughter cells that are genetically identical to each other and to the parent. OR The process of cell division in which a single cell produces two daughter cells which are identical to on another, and to the original parent cell.

Mle : Most Likely Estimate used by population geneticists. Maximum Likelihood Estimate used in BioGraphical Ancestry (BGA) tests.

Modal : The modal value is simply the most often occurring value or set of values in a given distribution of values or set of values. Compare it to the median value which has half of the distribution above it and half below it. The average or mean value is the sum of the values divided by the number of values. The average can be seriously skewed by a few very large or very small values. So, the modal haplotype is just the most frequently occurring haplotype in a set of haplotypes. (This definition is from John T. Nichols, May 2003, genealogydnal@rootsweb.com; permission granted.)

Molecular beacons : Real Time Amplification System.They are hairpin-shaped molecules that contain a stem-loop structure, a fluorophore and a quencher (Dabcyl). The stem sequence keeps the fluorophore and the quencher in close proximity so that any photons emitted by the fluorophore are absorbed by the quencher. The loop sequence is complementary with the target. When the probe finds its target, the loop opens and hybridises to the target. This removes the fluorophore from the vicinity of the quencher, allowing the fluorescence to appear and be measured. Molecular Beacons are designed to remain intact during the amplification reaction, and must rebind to target in every cycle for signal measurement. Molecular Beacons form a stem-loop structure when free in solution. Thus, the close proximity of the fluor and quench molecules prevents the probe from fluorescing. When a Molecular Beacon hybridises to a target, the fluorescent dye and quencher are separated, FRET does not occur, and the fluorescent dye emits light upon irradiation. Molecular Beacons can be used for multiplex assays by using spectrally separated fluor/quencher moieties on each probe. OR A molecular beacon is a single-strand oligonucleotide hybridization probe forming a stem-and-loop structure. The loop in this sequence is complementary to a target sequence on a larger nucleotide structure (DNA or RNA) and thus works as a probe, and the stem is formed by annealing the complementary arm sequences on either side of the loop. Molecular beacons do not fluorescence when free in solution, but when they are bonded to a target sequence, they undergo changes, primarily a separation of the stem sequences, that cause them to fluoresce brightly. This means that if the probe does not find a target, the molecular beacon will not glow.

Molecular biology : A modern branch of biology that attempts to explain biological phenomena in molecular terms. Most work in molecular biology is concerned with or uses genetic techniques. OR Study of the molecular basis of life including the biochemistry of molecules such as DNA RNA and proteins and the molecular structure and function of the various parts of living cells.

Molecular evolution : Study of how molecules have changed and evolved as a result of specialisation or selection.

Molecule : The smallest physical unit of a chemical compound, the defined collection of atoms bound together in a specific fashion. OR A group of atoms arranged to interact in a particular way; one molecule of any substance is the smallest physical unit of that particular substance.

Monoculture : Usually used in connection with crops, monoculture means the growing of identical plants in a large area, with no diversity. Natural meadows contain a mix of plant species, whereas a modern farmer's field is often a monoculture of just one species – and probably just one cultivar or variety of that species – grown at a high density.

Monomorphic probe : A DNA probe that will detect a DNA fragment that is the same size in everyone. Monomorphic probes are used to detect band shifting.

Multilocus probe : The original Jefferys process of simultaneously detecting up to ten or fifteen genetic loci simultaneously. Because of technical problems with this method it has been supplanted in forensic science by the single locus probe method.

Multiplex PCR : Multiplex PCR is the term used when more than one pair of primers is used in a PCR. The goal of multiplex PCR is to amplify several segments of target DNA simultaneously and thereby to conserve template DNA, save time, and minimize expense. It is a PCR strategy that enables the amplification of multiple DNA targets in one run. OR Multiplex PCR is simply a PCR reaction where two or more targets are detected in the same reaction. OR Multiplex PCR is a variation of PCR where multiple regions of DNA are amplified simultaneously.

Multiplex reaction : Used in the PCR method, it utilises a 'cocktail' of chemicals so that many different markers (or loci) may be copied and tagged at the same time.

Multiplexed : A system for analyzing several loci at once.

Multiplexing : A sequencing approach that uses several pooled samples simultaneously, greatly increasing sequencing speed.

Mutagen : An agent that gives rise to mutations. Common mutagens include UV radiation and xrays.

Mutation : A spontaneous or induced change in the DNA of a cell. OR Any inheritable change in DNA sequence. OR A change in the number, arrangement, or molecular sequence of a gene. OR Any inheritable changes in the DNA sequence that occurs during reproduction or cell division.

Mutation rate : The estimated rate of mutational change is estimated to be approximately one mutation per 500 transmission events (0.2%) for any given STR marker. Using 21 different markers, it is reasonable to expect one mutation every 24 transmission events. However, these changes are random and thus can take place at any time.

N

National dna index system (ndis) : Authorized by the DNA Identification Act of 1994, the FBI administers this national index. NDIS compares DNA profiles associated with a crime scene to DNA profiles collected from known convicted offenders, as well as to other crime scene profiles. When the DNA profiles are uploaded to NDIS, they are searched against the other DNA profiles submitted by other participating states.

Nested PCR : Nested PCR is a conventional PCR with a second round of amplification using a different set of primers. This second set of primers is specific to a sequence found within the DNA of the initial conventional PCR amplicon. The use of a second amplification step with the "nested" primer set results in a reduced background from products amplified during the initial PCR due to the nested primers' additional specificity to the region. The amount of amplicon produced is increased as a result of the second round of amplification and due to a reduction in any inhibitor concentrations. OR A very sensitive method for amplfication of DNA, which takes part of the product of a single PCR reaction (after 30-35 cycles), and subjects it to a new round of PCR using a different set of PCR primers which are nested within the region flanked by the original primer pair. ORA second PCR is performed on the product of an earlier PCR using primers which are internal to the originals. This improves sensitivity without impairing specificity. OR The second round amplification of an already PCR-amplified sequence using a new pair of primers which are internal to the original primers. Typically done when a single PCR reaction generates insufficient amounts of product.

Nitrogenous base : A molecule with the properties of a base, which also contains the element nitrogen.

Nonmatch : An individual is eliminated as the source of a biological sample. This occurs when one or more types from a specific location in the DNA of a known individual are not present in the type(s) for that specific location in the DNA obtained from an evidence sample.

Nuclear DNA : The DNA found in the nucleus of a cell.

Nucleated : A nucleus or occurring in the nucleus.

Nucleic acid : A large molecule composed of repeating nucleotides. OR DNA and RNA are both nucleic acids. They are long polymer structures made up of nucleotides. OR DNA or RNA. Each of these compounds consists of a backbone of sugar molecules ribose for RNA and deoxyribose for DNA linked by single phosphate groups. Attached to the sugars of the backbone are any of four nitrogenous bases, A, T, C or G for DNA and A, U, C or G for RNA. OR Nucleic acids are biochemical macromolecules that are composed of nucleotide chains, the genetic building blocks of life. DNA (deoxyribonucleic acid) and RNA (ribonucleic acid) are the most well known types of nucleic acid, and are found in living cells and viruses. OR A large molecule composed of nucleotide subunits.

Nucleic acid sequence based amplification (NASBA) : NASBA is an isothermal nucleic acid amplification procedure based on target-specific primers and probes, and the coordinated activity of THREE enzymes: AMV reverse transcriptase, RNase H and T7 RNA polymerase. NASBA allows direct detection of viral RNA by nucleic acid amplification.

Nucleolus : A dense, roughly spherical body found within the nucleus; the location in human cells where ribosomal RNA is transcribed.

Nucleotide : A subunit of DNA or RNA, consisting of one chemical base plus a phosphate molecule and a sugar molecule. OR A building block of DNA or RNA. OR A subunit of DNA or RNA, consisting of one chemical base plus a phosphate molecule and a sugar molecule. OR A molecular subunit of DNA or RNA consisti ng of a base (adenine, guanine, thymine, or cytosine in DNA; adenine, guanine, uracil, or cytosine in RNA). Thousands of nucleotides are linked to form a DNA or RNA molecule. See DNA, base pair, RNA.

Nucleus : The cell structure that houses the chromosomes. OR The organelle in plant and animal cells which houses the DNA. OR The central structure of a cell that contains the genetic material. OR A cell organelle in all cell types except bacteria; inside of the organelle the chromosomes are contained.

O

Obligate gene : In parentage testing the gene that must have been passed by the father to the child.

Off ladder alleles : An allele that is not included in the allelic ladder. These can be true, rare alleles or artifacts.

Okzaki fragments : The fragments of DNA synthesized on the lagging strand during DNA replication Named after their discoverer, Japanese biochemist Reiji Okazaki.

Oligonucleotide : A short segment of DNA that is synthesised artificially. OR A molecule usually composed of 25 or fewer nucleotides; used as a DNA synthesis primer.

Oncogenes : Genes that normally play a role in the growth of cells but, when overexpressed or mutated, can foster the growth of cancer.

Organism : Any living thing.

Origins of replication : The nucleotide sequence at which DNA synthesis is initiated.

P

Partial profile : DNA evidence that does not yield identifiable results in all 13 core loci.

PCR (Polymerase chain reaction) : Polymerase chain reaction. A method of amplifying a defined region or target of dna by repeated cycles of denaturation (heat), annealing and extension. After 30 cycles, a single target dna can be amplified over a billion fold. OR Polymerase chain reaction A process carried out in a test tube that produces millions of copies of small sections of DNA. Using a heat resistant enzyme (DNA polymerase) and a mixture of other chemicals, cycles of hot and cold temperatures essentially photocopy a particular marker (or locus) of the DNA many times. Fluorescent tags are added to each copy so that they may be detected using laser analysers. A technique called multiplexing enhances the process. OR The polymerase chain reaction is a technique for quickly "amplifying" a particular piece of DNA in the test tube (rather than in living cells). Thanks to this procedure, it's possible to make virtually unlimited copies of a single DNA molecule even though this molecule may be present in a mixture containing many other different DNA molecules. In order to perform PCR, you must know at least a portion of the sequence of the DNA molecule that you wish to replicate. OR A "biological copy machine" a method for making many copies of a specific DNA base sequence. OR A process used in DNA identification testing in which one or more specific small regions of the DNA are copied using a DNA polymerase enzyme so that a sufficient amount of DNA is generated for analysis.

PCR additives : PCR additives, including formamide, DMSO, glycerol, betaine, and PCRx Enhancer Solution can enhance amplification. Their proposed mechanism is to lower the melting temperature, thereby aiding primer annealing and helping the DNA polymerase extend through regions of secondary structure. OR Amplification of a DNA target by the polymerase chain reaction (PCR) often requires laborious optimization efforts. In this regard, the use of certain organic chemicals such as dimethyl sulfoxide, polyethylene glycol, betaine and formamide as cosolvents has been found to be very helpful. Unfortunately, very little is known about the precise structural features that make these additives effective and, accordingly, the number of such chemicals currently known to enhance PCR is limited

Pcr inhibitors : A substance that interferes with the Polymerase Chain Reaction (PCR) process. Examples of PCR inhibitors include dyes, soil, chemicals, and heme (hemoglobin).

Pcr-elisa: A capture assay for nucleic acids that mimic enzyme linked immunosorbant assays. In this assay, PCR products hybridized to an immobilized capture probe. OR PCR-ELISA is PCR that is followed by capture and hybridization in microtiter plates to labeled probes and detection similar to ELISA.

Pcr-rflp (polymerase chain reaction-restriction fragment length polymorphism) : Alternative denomination: cleaved amplified polymorphic sequence Following PCR amplification of a locus, the amplicon is treated with a restriction endonuclease. If the recognition site for this enzyme is present in the amplicon, two or more restriction fragments are generated. Thus sequence variation between individuals at the recognition site(s) can be detected by electrophoresis.

PCR-SSCP (Polymerase chain reaction single-strand conformation polymorphism): PCR-SSCP is a simple procedure where denatured PCR products are electrophoresed through a non-denaturing polyacrylamide gel. The single strands adopt primary conformations that are dependent on their nucleotide sequence and this determines the rate at which they migrate through the gel matrix. Each PCR product with a different sequence therefore, will be theoretically represented by two bands corresponding to the two strands of the amplified molecule

Pentose suga : A sugar molecule containing five carbon atoms. E.g. deoxyribose.

Pfu DNA polymerase : Pfu DNA polymerase is an enzyme found in the hyperthermophilic archaeon Pyrococcus furiosus, where it functions in vivo to replicate the organism's DNA. In vitro, Pfu is used to quickly amplify DNA in the Polymerase Chain Reaction, where the enzyme serves the central function of copying a new strand of DNA during each extension step.The main difference between Pfu and alternative enzymes is Pfu's superior thermostability and 'proofreading' properties compared to other thermostable polymerases.

Phenotype : The outward physical characteristics of an organism.

Physical maps : DNA maps showing the location of identifiable landmarks, either genes or distinctive short sequences of DNA. The lowest resolution physical map shows the banding pattern on the 24 different chromosomes; the highest resolution map depicts the complete nucleotide sequence of the chromosomes. (See Contig maps.)

Point mutation : A mutation that can be mapped to one specific site within a locus. A small mutation that consists of the replacement (transition or transversion); addition; or deletion (frameshift) of one or a few bases. OR An alteration in DNA sequence caused by a single nucleotide base change, insertion, or deletion. OR A change in a single base pair of DNA.

Polya Tail : A long series of adenines added to the 3' end of messenger RNA that allows the molecule to travel out of the nucleus and helps keep track of the number of translations the molecule can undergo before is begins to fall apart.

Polymer : A large molecule formed by the combination of many similar smaller molecules called monomers.

Polymerase (DNA) : An enzyme which assembles nucleic acid residues into DNA or RNA polymers. Polymerases work from the DNA complement of the sequence to be built. DNA polymerases copy DNA to DNA to replicate the genome before mitosis, while RNA polymerases copy DNA to RNA as the first step in gene transcription. OR General term for enzymes that carry out the synthesis of nucleic acid, using a pre-existing nucleic acid template & appropriate nucleotides (viz. ribonucleotides for RNA and deoxyribonucleotides for DNA). OR A polymerase is a naturally occurring enzyme, a biological macromolecule that catalyzes the formation and repair of DNA (and RNA). The accurate replication of all living matter depends on this activity.

Polymorphic : A highly variable genetic trait in a population. An rflp locus that is polymorphic will have many possible size variants between individuals.

Polymorphic locus : A genetic locus for which there are many alleles in existence. The loci used in forensic science are the most polymorphic known.

Polymorphism : A polymorphism is a DNA sequence variation that is common in the population. The arbitrary cut-off point between a mutation and a polymorphism is 1 per cent. That is, to be classed as a polymorphism, the least common allele must have a frequency of 1per cent or more in the population. If the frequency is lower that this, the allele is regarded as a mutation. OR Differences in DNA sequences that occur naturally in a population. Single nucleotide substitutions, insertions and deletions of nucleotides and repetitive sequences (microsatellites) are all examples of a polymorphism. The position at which such a sequence difference is found is a polymorphic site. OR A variant form of a gene. Most polymorphisms are harmless and are part of normal human genetic variation.

Polypeptide : A molecule made up of a string of amino acids. A protein is an example of a polypeptide.

Population : A group of individuals residing in a given area at a given time. OR A defined group of individuals for which databases of gene frequencies are collected by sampling members of the population.

Posttranscriptional processing : The alteration of the RNA transcript of the DNA. This may involving cutting and splicing RNA, or modifying it in other ways.

Predictive gene tests : Tests to identify gene abnormalities that may make a person susceptible to certain diseases or disorders.

Primary lesion : changes in DNA caused by mutagens.

Primase : An enzyme that catalyzes the construction of the short RNA nucleotide polymers used as primers by DNA polymerases.

Primers : PCR primers are short fragments of single stranded DNA (15-30 nucleotides in length) that are complementary to DNA sequences that flank the target region of interest. The purpose of PCR primers is to provide a "free" 3′-OH group to which the DNA polymerase can add dNTPs. OR A primer is a short segment of nucleotides which is complementary to a section of the DNA which is to be amplified in the PCR reaction. Primers are annealed to the denatured DNA template to provide an initiation site for the elongation of the new DNA molecule. OR A short piece of DNA, usually synthetic, that defines the specific site on a DNA molecule for a DNA polymerase to start making new DNA. An essential ingredient in the PCR reaction mix.

Probability of Exclusion : The probability that a random individual would be excluded as the source of analyzed DNA evidence.

Probability of inclusion : The probability that a random individual would be included as a potential source of analyzed DNA evidence.

Probe : Is a term to describe a reagent used to make a single measurement in a gene expression experiment.

Probe : Single-stranded DNA or RNA molecules of specific base sequence, labeled either radioactively or immunologically, that are used to detect the complementary base sequence by hybridization. OR Defined nucleic acid (DNA or RNA) that can be used to identify, usually through autoradiography, specific DNA or RNA molecules bearing the complementary sequence. A labeled (radioactive; antigen; enzyme; fluorescent) nucleic acid complementary to the sequence being searched for in a restriction digest, genome library, northern blot or in situ hybridization. OR A finite nucleic acid piece that can be used to identify specific DNA segments bearing its complementary sequence.

Profile : A discription of an individual or evidence resulting from the examination of one or more polymorphic loci

Promoter : the portion of DNA within the regulatory region that directs the binding of RNA polymerase and the transcription of a coding region which follows.

Proofreading : In DNA synthesis, the process of recognizing a basepair error during the polymerization events and correcting it. OR Fidelity of DNA replication in vivo is achieved, in part, by the proof-reading activity of the DNA polymerase. When an errant nucleotide is incorporated and forms a mismatch with the template, it is removed by a 3′ to 5′ exonuclease proof-reading activity associated with the polymerase. Some thermostable DNA polymerases have proof-reading activity, a characteristic desirable for accurate DNA amplification and for PCR amplification of long DNA sequences. The thermostable DNA polymerase most widely used for PCR is Taq polymerase, which lacks proof-reading activity.

Protein : A type of molecule occurring in all living things. Proteins are made from about 20 basic units (See amino acid). There is a huge variation in protein size and function depending on how the units are put together. The instructions for how to assemble proteins are usually contained within DNA molecules. Within cells, proteins carry out most of the chemical functions necessary for life - for example, building other proteins, carrying out chemical reactions, controlling what enters and leaves the cell, making structures, controlling the expression of genes.

Proteinase K : An enzyme that digests protein. Used to isolate dna from cells or tissue.

Proteomics/ proteome : All the protein products produced by an organism's genes are referred to as the proteome (by analogy with the genome). Proteomics is the study of these proteins and how they interact to affect the life of the organism.

Protocol : Laboratory procedures manual.

Pullup : An artifact resulting from an intense peak in one color channel, which is detected as a minor peak in another color channel. OR Specifically related to ABI Gene Scan® software, a peak seen in one color that is not due to the presence of DNA, but to incorrect compensation for the spectral overlap of the four dyes used in detecting multiple loci in one reaction. See also Artifact

Purine : A nitrogen containing, single ring compound that occurs in nucleic acids. In DNA molecules, the purines are adenine and guanine.

Pyrimidibe dimer : primary lesions caused by UV radiation.

Pyrimidine : A nitrogen containing, double ring compound that occurs in nucleic acids. In DNA molecules, the pyrimidines are cytosine and thymine.

Q

Quality assurance (qa) : A program conducted by a laboratory to ensure accuracy and reliability of tests performed. OR Externally administered programs that are design to assess and maintain a minimum level of laboratory performance. It includes laboratory inspection and accreditation, personnel training, examination and certification, regular proficiency testing and the maintenance of a quality control program.

Quality assurance standards : The standards originated from the DNA Advisory Board. They place specific requirements on labs involved in forensic DNA analysis.

Quality control : Systematic regular activities for verifying and maintaining a specified level of quality in a product or process. It includes careful planning, regular documentation, continued inspection and measurement and the implementation of corrective action when necessary.

Quantitative PCR : A PCR method that enables the number of DNA molecules in a sample to be estimated.

Quencher : A molecular entity that deactivates (quenches) an excited state of another molecular entity, either by energy transfer, electron transfer, or by a chemical mechanism. OR The molecule that absorbs the emission of fluorescent reporter when in close vicinity.

Quenching : The reduction of luminescence of a sample by the addition of a quencher.

R

Race : (In genetics) a population or group that has a characteristic set of gene frequencies

RACE-PCR (rapid amplification of cdna ends) : RACE-PCR is an anchor PCR modification of RT-PCR. The rationale is to amplify sequences between a single previously characterized region in the mRNA (cDNA) and an anchor sequence that is coupled to the 5 or the 3 end. A primer is designed from the known internal sequence and the second primer is selected from the relevant anchor sequence.

Random match probability : The probability that the DNA in a random sample from the population has the same profile as the DNA in the evidence sample.

RAPD (Randomly amplified polymorphic DNA) : A technique for amplifying anonymous stretches of DNA, using PCR with arbitrary primers.

RE test gel : A gel used to analyze DNA that has been cut by a restriction enzyme (RE) to ensure that the cutting of the DNA has been complete.

Sometimes contaminants in forensic samples will inhibit the action of the enzyme and prevent it from cutting the DNA.

Real Time PCR (Originally described as "Kinetic PCR" by Higuchi et al. In 1993) : Real-time PCR is so named because it detects and measures the amplification of target nucleic acids as they are produced. Real-time PCR requires the use of primers similar to those used in conventional PCR. However, unlike conventional PCR, real-time PCR uses an oligonucleotide probe labeled with fluorescent dyes or an alternative fluorescent detection chemistry, and a thermocycler equipped with the ability to measure fluorescence. OR Real-Time PCR, also called quantitative (real-time) PCR or Real-Time Quantitative PCR (RTQ-PCR), is a method of simultaneous DNA quantification and amplification. DNA is specifically amplified by polymerase chain reaction. After each round of amplification, the DNA is quantified. Common methods of quantification include the use of fluorescent dyes that intercalate with double-strand DNA and modified DNA oligonucleotides (called probes) that fluoresce when hybridized with a complementary DNA. OR A technique designed to detect and quantify sequence-specific PCR products as they accumulate in 'real-time' during the PCR amplification process.

Recessive allele : A gene that is expressed only when its counterpart allele on the matching chromosome is also recessive (not dominant). Autosomal recessive disorders develop in persons who receive two copies of the mutant gene, one from each parent who is a carrier.

Recessive gene : A gene which must b e present on both chromosomes in a pair to show outward signs of a certain characteristic.

Recombinant DNA technology : The techniques and tools employed to produce recombinant DNA. OR A form of DNA produced by splicing together segments of DNA from two or more organisms.

Recombination : The reversal of coupling phase I meiosis as gauged by the resulting phenotype.

Region : A sequence found within the promoter of some prokaryotic genes about 30 to 35 base pairs upstream of the transcription initiation site.

Regulatory region : The portion of a DNA molecule that is responsible for regulating transcription of the coding region.

Replication : The enzymatic process of making new DNA based on the use of use of an existing strand of DNA as a template. The enzymes that perform this process require a primer, a short piece of DNA complementary to the template that defines the base at which the replication will start and that will become part of the newly synthesized DNA. restriction enzymes (RE), restriction endonucleases. A class of enzymes obtained from microorganisms that cut the DNA strands at specific four to twelve base pair sequences. There are over one hundred restriction enzymes commercially available each having a different recognition site. In forensic science the most commonly used RE is called Hae III although there are some labs who still use Hinf I or Pst I.

Rep-PCR : Is a type of polymerase chain reaction that targets the repetitive sequences in bacterial genomes using specific primers that are designed complementary to bacterial interspersed repetitive sequences.

Repression : Is decreasing the rate of gene expression.

Reproductive cells : Egg and sperm cells. Each mature repro-ductive cell carries a single set of 23 chromosomes.

Restriction enzyme : These are enzymes (endonucleases, more specifically) which recognize a specific, short sequence of DNA and cut the DNA at that point. Different restriction enzymes recognize and cut different sequences. There are hundreds of different restriction enzymes available commercially. Many restriction enzymes leave "sticky ends" when they cut, which are available to bind with other "sticky ends" left by the same enzymer. Restriction enzymes are a vital tool in genetics, since they allow cutting (and pasting) of DNA. OR A protein that recognizes specific, short nucleotide sequences and cuts DNA at those sites. Bacteria contain over 400 such enzymes that recognize and cut more than 100 different DNA sequences. OR An endonuclease that will recognise a specific target sequence and cut the DNA chain at that point.

Restriction enzyme cutting site : A specific sequence of DNA at which a particular restriction enzyme cuts the DNA.

Restriction enzyme, endonuclease : A protein that recognizes specific, short sequences of DNA and cuts at those sites. See also Restriction enzyme cutting site.

Restriction fragment length polymorphism (RFLP) : Variations in the length of a DNA restriction fragment caused by the presence of a polymorphic locus within the fragment. Analysis of RFLPs by the Southern blotting method is the most common and definitive forensic DNA test.

Restriction site : The specific nucleotide sequence of DNA at which a particular restriction enzyme cuts the DNA. A sequence of DNA that is recognized by an endonuclease (a protein that cuts DNA) as a site at which the DNA is to be cut.

Reverse transcriptase : RNA-dependent DNA polymerase An enzyme that uses an RNA molecule as a template for the synthesis of a complementary DNA (cDNA) strand. OR A reverse transcriptase, also known as RNA-directed DNA polymerase, is a DNA polymerase enzyme that transcribes single-stranded RNA into double-stranded DNA. Normal transcription involves the synthesis of RNA from DNA, hence reverse transcription is the reverse of this, as it synthesises DNA from RNA. OR An enzyme used by retroviruses to form a complementary DNA sequence (cDNA) from their RNA. The resulting DNA is then inserted into the chromosome of the host cell.

Reverse transcription : The copying of an RNA molecule back into its DNA complement. The enzymes that perform this function are called reverse transcriptases. Reverse transcription is is used naturally by retroviruses to insert themselves into an organism's genome. Artifically-induced reverse transcription is a useful technique for translating unstable mRNA molecules into stable cDNA. OR The process of copying information found in RNA into DNA.

Ribonucleic acid (RNA) : A chemical cousin of DNA, RNA (ribonucleic acid) is responsible for translating the genetic code of DNA into proteins.

Ribosomal RNA (rrna) : A class of RNA found in the ribosomes of cells. **Ribosome :** Organelle of the cell that acts as a site of protein synthesis. OR The small cellular structure in which RNA translates the genetic code into proteins.

RNA (ribonucleic acid) : A chemical similar to DNA. The several classes of RNA molecules play important roles in protein synthesis and other cell activities. OR A messenger molecule which copies the information from DNA (which is housed in the nucleus), and transports the code outside the nucleus to cellular machinery which reads the code and puts together a long chain of amino acids (called a "protein"). The RNA molecule is very similar to DNA.

Rna polymerase : The enzyme of RNA transcription.

RNA ribonucleic acid : A nucleic acid polymer similar to DNA and used to translate the genetic code into a form useful by the cells.

S

Satellite DNA : Satellite DNA consists of highly repetitive DNA, and is so called because repetitive DNA sequences tend to have a relatively high frequency of the nucleotides Adenine and Thymine, and thus have lower density - such that they form a second 'satellite' band when genomic DNA is separated along a density gradient. OR A portion of the DNA that differs enough in base composition so that it forms a distinct band on cesium chloride gradient centrifugation; usually contains highly repetitive DNA sequences. OR DNA that forms a separate band in a bouyant density gradient because of its different nucleotide composition (A:T rich DNAs are less dense than G:C rich DNAs). Highly repetitive eukaryotic DNA primarily located around centromeres. Satellite DNA usually has a different buoyant density than the rest of the cell's DNA.

SCAR (Sequence characterised amplified region) : A locus representing a single RAPD fragment which has been sequenced. Primers specific to the locus can be designed and used in PCR amplification.

Screening : Looking for evidence of a particular disease such as cancer in persons with no symptoms of disease.

Sdis : State DNA Index System containing the statelevel DNA records uploaded from local laboratory sites within the state. SDIS is the state's repository of DNA identification records and is under the control of state authorities. The SDIS laboratory serves as the central point of contact for access to the National DNA Index System (NDIS). The DNA Analysis Unit I (DNAUI) serves as the SDIS laboratory for the FBI.

Secretory vesicle : A vesicle carrying proteins destined for secretion. Secretory vesicles bud off the Golgi apparatus and travel to the plasma membrane where the vesicle contents are released by exocytosis.

Selectin : A membrane glycoprotein containing a lectinlike domain that binds oligosaccharide groups. is required for selectin binding to oligosaccharides.

Self-splicing intron : An intron that is excised in a reaction mediated by the RNA precursor itself.

Semiconservative replication : The mode of duplicating DNA in which each strand serves as a template for the synthesis of a complementary strand. The result is two molecules of doublestranded DNA, each of which contains one of the parental strands.

Sense strand : In double-stranded DNA the sense strand is the strand that contains codons. Also called the coding strand. The opposite strand is called the antisense strand or the template strand.

Sequence (or site) heteroplasmy : Presence of mtDNA molecules that have different nucleotides at the same address.

Sequencer : An apparatus used for deciphering the order of bases in a stra nd of DNA.

Sequential reaction : An enzymatic reaction in which all the substrates must be bound to the enzyme before any product is released.

Serine protease : A protease with an active-site serine residue that acts as a nucleophile during catalysis.

Shine-Dalgarno
sequence : A purine-rich region just upstream of the initiation codon in prokaryotic mRNA molecules. The Shine-Dalgarno sequence binds to a pyrimidine-rich sequence in the ribosomal RNA, thereby positioning the ribosome at the initiation codon.

Short tandem repeat (str): a short pattern (often two to five bases in length) *repeated* a number of times in a row (in *tandem*); for instance, GATAGATAGATA, three repeats of the GATA sequence; the differences in the STRs at selected markers on the Y chromosome provide a basis for comparison among individuals and populations and are used extensively for most YDNA genetealogical testing; also called a microsatellite (From glossary in Trace Your Roots With DNA by Smolenyak and Turner). OR DNA analysis method which targets regions on the chromosome which contain multiple copies of an identical DNA sequence in succession. OR Multiple copies of a short identical DNA sequence arranged in direct succession in particular regions of chromosomes.

Shuttle vector : A cloning vector that can replicate in both prokaryotic and eukaryotic cells. Shuttle vectors are used to transfer recombinant Ca2+ DNA molecules between prokaryotic and eukaryotic cells.

Sicklecell anemia : An inherited, potentially lethal disease in which a defect in hemoglobin, the oxygencarrying pigment in the blood, causes distortion (sickling) and loss of red blood cells, producing damage to organs throughout the body. OR disease characterized by the presence of defective hemoglobin molecules that cause red blood cells to be malformed, often taking on a sickle shape. The malformed cells can't carry oxygen properly, and can tend to clog capillaries and small veins. Symptoms include joint pain, acute abdominal pain, and ulcerations of the lower extremities.

Signal transduction : The process whereby an extracellular signal is converted to an intracellular signal by the action of a membrane-associated receptor, a transducer, and an effector enzyme.

Signal-recognition particle (SRP) : A eukaryotic protein-RNA complex that binds a newly synthesized peptide as it is extruded from the ribosome. The signal-recognition particle is involved in anchoring the ribosome to the cytosolic face of the endoplasmic reticulum so that protein translocation to the lumen can occur.

Single-strand binding protein (SSB) : A protein that binds tightly to single-stranded DNA, preventing the DNA from folding back on itself to form double-stranded regions.

Site-directed mutagenesis : An in vitro procedure by which one particular nucleotide residue in a gene is replaced by another, resulting in production of an altered protein sequence.

Small nuclear ribonucleoprotein (snrnp) : An RNA-protein complex composed of one or two specific snRNA molecules plus a number of proteins. snRNPs are involved in splicing mRNA precursors and in other cellular events.

Small RNA : A class of RNA molecules. Some small RNA molecules have catalytic activity. Some small nuclear RNA molecules (snRNA) are components of small nuclear ribonucleoproteins (snRNPs).

Solvation sphere : The shell of solvent molecules that surrounds an ion or solute.

Solvation : A state in which a molecule or ion is surrounded by solvent molecules.

Special pair : A specialized pair of chlorophyll molecules in reaction centers that is the primary electron donor during the light-dependent reactions of photosynthesis.

Splice site : The conserved nucleotide sequence surrounding an exon-intron junction. It includes the site where the RNA molecule is cleaved during intron excision.

Spliceosome : The large protein-RNA complex that catalyzes the removal of introns from mRNA precursors. The spliceosome is composed of small nuclear ribonucleo-proteins.

Splicing : The process of removing introns and joining exons to form a continuous RNA molecule.

Stacking interactions : The weak noncovalent forces between adjacent bases or base pairs in single-stranded or double-stranded nucleic acids, respectively. Stacking interactions contribute to the helical shape of nucleic acids.

Standard free-energy change : The freeenergy change for a reaction under biochemical standard-state conditions.

Sticky end : An end of a double-stranded DNA molecule with a single-stranded extension of several nucleotides.

Strand invasion : The exchange of single strands of DNA from two nicked molecules having homologous nucleotide sequences.

Subcloning : Transferring cloned DNA between cloning vectors.

Substrate cycle : A pair of opposing, metabolically irreversible reactions that catalyzes a cycle between two pathway intermediates. Substrate cycles provide sensitive regulatory sites.

Substrate : A reactant in a chemical reaction. In enzymatic reactions, substrates are specifically acted upon by enzymes, which catalyze the conversion of substrates to products.

Substrate-level phosphorylation : Phosphorylation of a nucleoside diphosphate by transfer of a phosphoryl group from a non-nucleotide substrate.

Subunit (sigma subunit) : A subunit of prokaryotic RNA polymerase, which acts as a

Supercoil : A topological arrangement assumed by over- or underwound double-stranded DNA. Underwinding gives rise to negative supercoils; overwinding produces positive supercoils.

Svedberg unit (S) : A unit of second used for expressing the sedimentation coefficient, a measure of the rate at which a large molecule or particle sediments in an ultracentrifuge. Large S values usually indicate large masses.

T

TATA box : An A/T-rich DNA sequence found within the promoter of both prokaryotic and eukaryotic genes.

Template strand : The strand of DNA within a gene whose nucleotide sequence is complementary to that of the transcribed RNA. During transcription, RNA polymerase binds to and moves along the template strand in the direction, catalyzing the synthesis of RNA in the direction.

Termination codon : A codon that is recognized by specific proteins that cause newly synthesized peptides to be released from the translation machinery thus terminating translation. The three termination codons (UAG, UAA, and UGA) are also known as stop codons.

Tertiary structure : The compacting of polymeric chains into one or more domains within a macromolecule.In proteins, tertiary structure is stabilized mainly by hydrophobic interactions between side chains.

Topoisomerase : An enzyme that alters the supercoiling of a DNA molecule by cleaving a phosphodiester linkage in either one or both strands, rewinding the DNA, and resealing the break. Some topoisomerases are also known as DNA gyrases.

Transcription : The copying of biological information from a double-stranded DNA molecule to a single-stranded RNA molecule, catalyzed by a transcription complex consisting of RNA polymerase and associated factors.

Transcription bubble : A short region of doublestranded DNA that is unwound by RNA polymerase during transcription.

Transcription factor : A protein that binds to the promoter region, to RNA polymerase, or to both during assembly of the transcription initiation complex. Some transcription factors remain bound during RNA chain elongation.

Transcription initiation complex : The complex of RNA polymerase and other factors that assembles at the promoter at the start of transcription.

Transcriptional activator : A regulatory DNAbinding protein that enhances the rate of transcription by increasing the activity of RNA polymerase at specific promoters.

Transfection : The introduction of foreign DNA into a cell via a virus or phage vector.

Transferase : An enzyme that catalyzes a grouptransfer reaction. Transferases often require a coenzyme.

Transgenic organism : An individual organism that carries recombinant DNA stably integrated in all of its cells.

Translation complex : The complex of a ribosome and protein factors that carries out the translation of mRNA in vivo.

Translation initiation complex : The complex of ribosomal subunits, an mRNA template, an initiator tRNA molecule, and initiation factors that assembles at the start of protein synthesis.

Translation initiation factor : A protein involved in the formation of the initiation complex at the start of protein synthesis.

Translation : The synthesis of a polypeptide whose sequence reflects the nucleotide sequence of an mRNA molecule. Amino acids are donated by activated tRNA molecules, and peptide bond synthesis is catalyzed by the translation complex, which includes the ribosome and other factors.

Translational frameshifting : The shift in reading frame that may occur during the translation of an mRNA molecule. The synthesis of some proteins requires translational frameshifting.

Transposon : A mobile genctic element that jumps between chromosomes or parts of a chromosome by taking advantage of recombination mechanisms. Also known as a transposable element or an integron.

Transverse diffusion : The passage of lipid or protein molecules from one leaflet of a lipid bilayer to the other leaflet. Unlike lateral diffusion within one leaflet of a bilayer, transverse diffusion is extremely slow.

Trna : A class of RNA molecules that carry activated amino acids to the site of protein synthesis for incorporation into growing peptide chains. tRNA molecules contain an anticodon that recognizes a complementary codon in mRNA.

U

Uncoupling agent : A compound that disrupts the usual tight coupling between electron transport and phosphorylation of ADP.

Uniport : The transport of a single type of solute across a membrane by a transport protein.

V

Vacuole : A fluid-filled organelle in plant cells that is a storage site for water, ions, or nutrients.

Van der Waals force : A weak intermolecular force produced between neutral atoms by transient electrostatic interactions. Van der Waals attraction is strongest when atoms are separated by the sum of their van der Waals radii; strong van der Waals repulsion precludes closer approach.

Van der Waals radius : The effective size of an atom. The distance between the nuclei of two nonbonded atoms at the point of maximal attraction is the sum of their van der Waals radii.

Variable arm : The arm of a tRNA molecule that is located between the anticodon arm and the arm. The variable arm can range in length from about 3 to 21 nucleotides.

Virus : A nucleic acid–protein complex that is capable of invading a host cell. A virus takes over the transcription and replication machinery of the host cell to replicate itself, using both its own and the host's gene products.

Vitamin : An organic micronutrient that cannot be synthesized by an animal and must be obtained in the diet. Many coenzymes are derived from vitamins.

W

Wobble position : The position of an anticodon, where non-Watson-Crick base pairing with a nucleotide in mRNA is permitted. The wobble position makes it possible for a tRNA molecule to recognize more than one codon.

X

X-ray crystallography : A technique used to determine secondary, tertiary, and quaternary structures of biological macromolecules. In X-ray crystallography, a crystal of the macromolecule is bombarded with X rays, which are diffracted and then detected electronically or on a film. The atomic structure is deduced by mathematical analysis of the diffraction pattern.

Y

Yeast artificial chromosome (YAC) : A cloning vector that contains yeast centromeric DNA and a yeast origin of replication. A YAC can accommodate up to 500 kilobase pairs of insert DNA. The artificial chromosome can be introduced into a yeast cell, where it behaves like a normal chromosome during replication.

Z

Z-dna : A conformation of oligonucleotide sequences containing alternating deoxycytidylate and deoxyguanylate residues. Z-DNA is a lefthanded double helix containing approximately 12 base pairs per turn.

Zwitterion : A molecule containing negatively and positively charged groups.

INDEX

A

B

C

□□□